The Blacksmith's Craft

THE
Blacksmith's
CRAFT

A PRIMER OF TOOLS AND METHODS

Charles McRaven

Storey Publishing

To Bill Cameron, the best blacksmith's helper I ever had.

*The mission of Storey Publishing is to serve our customers by
publishing practical information that encourages
personal independence in harmony with the environment.*

Edited by Carleen Perkins
Cover design by Kent Lew
Text design and production by Ron Toelke, Toelke Associates
Line drawings by Chandis Ingenthron and Charles McRaven
Photography by Linda Moore McRaven, Charles McRaven, and Ashley McRaven
Indexed by Christine R. Lindemer, Boston Road Communications

Printed in the United States by Versa Press
20 19 18 17 16 15 14 13 12 11

Library of Congress Cataloging-in-Publication Data

McRaven, Charles.
    The blacksmith's craft : a primer of tools and methods / Charles McRaven.
        p. cm.
    Includes bibliographical references and index.
    ISBN 978-1-58017-593-7 (pbk. : alk. paper)
    1. Blacksmithing. I. Title.
TT220.M36 2005
682—dc22
                                                                        2005004013

TABLE OF CONTENTS

PREFACE TO THE NEW EDITION

With the back-to-the-land movement largely a memory, the role of blacksmithing as a craft has changed. In the early 1980s, when this book was first published as *Country Blacksmithing,* homesteaders learned crafts like blacksmithing to assert their independence from the modern world of factory-made throwaways. They wanted to create something that was beautiful, useful, and home-fashioned.

Although most blacksmiths today tend to be hobbyists, not homesteaders, their eagerness to learn blacksmithing comes from the same desire for authenticity. Like most craftspeople, the smith produces one-of-a-kind creations that people want more than they may really need. Whether forging fine knives of carbon or Damascus steel or creating delicately wrought iron gates, the smith's work today is cultural rather than purely necessary.

There are those who make a living from it. In our area are two highly successful smiths who cater to wealthy clients. Both specialize in grilles, gates, and other ornamental iron. Both have spent years building names for themselves, and their work is now in demand. One has grillwork in the National Cathedral in Washington, D.C.

Most of us, though, practice our craft not as our principal livelihoods. I know a carpenter who does smithing as a sideline. He can forge a latch or a set of hinges at a reasonable price. A stone mason I know once thought he could stay alive by smithing. He couldn't, but he still enjoys it, specializing in tool forging. My friend David Morris, a doctor, has a complete shop at his home. I met him when he took a six-day workshop from me to learn how to use all those tools.

I used to go out of my way to fashion necessities of iron and steel myself, scorning factory-made products I didn't have the money to buy anyway. Today, however, my blacksmithing is mostly limited to the needs of the historic buildings we restore, and an occasional commission. I still forge knives, chisels, pot racks, fireplace cranes, and andirons, but more often ornamental hinges, Norfolk or Suffolk latches, reproduction sconces, and other wrought-iron fixtures for whatever house we're working on at the time.

But aside from my restoration work, I do smithing because I enjoy it. I like being able to forge anything I can imagine and anything I might need, from stray pieces of iron or steel. And I know each piece will be the only one just like that. Your forged work will become heirlooms.

The decision to forge something yourself often comes down to whether you want the hand-forged chisel or you'd be happy with the plastic-handled one from the hardware store. Case in point — my daughter Chelsea was studying geology in college. She wanted a geologist's stone hammer, and she didn't want a factory-made one. She didn't ask *if* I could forge it for her, but *how soon* could I do it?

Well, now. How do you turn down such an opportunity? I had some good coil spring stock in the shop, after all. Took just over half a day to make it, with polished ash grips and the tempering colors left in. No way could a mass-produced hammer have filled the bill nearly so well. Chelsea loans it only to special people who really appreciate it.

You might also find yourself forging something because you just can't get it anywhere else. When I bought an old Case tractor a few years ago, I designed and built a three-point hitch for it. Then I discovered the hydraulic pump shaft, driven off the engine timing gear, was broken. No replacement available. My choices were either to have a machine shop duplicate the shaft or do it myself. The shaft was threaded at one end, with a step to a smaller diameter. It wouldn't be simple, but hey, I'm a blacksmith, right? The shaft wasn't perfect, but it's still functioning years later.

Be warned, though. This craft will get into your blood. No matter how serious you are about blacksmithing, whether you're a purist or a hobbyist, a beginner or a veteran, you will never again look at the black metal in the same way. And it'll be hard to leave it alone.

Charles McRaven
Albemarle County, Virginia
January 2005

1

THE PIONEER SMITH

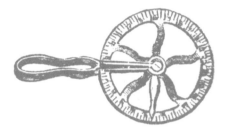

Setting up your forge on your own farm or modern home-stead will parallel so nearly the experience of the pioneer blacksmith coming into the new country that comparisons are inevitable. You may already have your place established — cabin, clearing, garden — and you will no doubt be a settler first and a blacksmith second, instead of the other way around. You may wind along your woods road in a wheezing pickup truck or a station wagon with your anvil sitting solid back there and your tools clinking on the way home. But it's all so nearly the same. The early smith made settlement possible; in this day the craft can make it so much better, so much nearer the independent ideal.

The pioneer smith came into the new country much like the other early settlers did — with axe and rifle and his family, if he had one, in the search for land. He was most often a farmer, too, and the hunger for new territory drove him, as it drove the others, west from the Carolinas, from Virginia, and from the Northeast.

But this man had in his wagon a treasured anvil, a hammer or two, an assortment of scraps and bits of iron, and a head full of answers to the mysteries of working the black metal. He'd served a long apprenticeship in a shop in the eastern settlements, and maybe as a journeyman in a wagon works or a musket factory.

And his wagon, laden with the barest needs for a frontier home, moved slowly with its weight of iron toward the beginnings of a settlement. A primitive gristmill might be set up at a big spring, its timbers raw and new among the fresh stumps and the falling water. These two craftsmen often located together, the smith and the miller, to serve the scattering of backwoods cabins hewn from the crowding forests.

The smith would find a welcome among the hunters and the settlers in the creekside clearings. He would often be taken into an already crowded cabin with his family so he could get his shop put up sooner. His house could come later, with a community log-raising marked by hard work, good food, courting, games, and raw, new whiskey from yet another craftsman's still.

Blacksmithing has changed little since the pioneer smith set up his forge in the new country.

Under a shelter of some kind, the smith would build his forge, often of stones laid up with clay between, with a passage for air from the giant wood-and-deerskin bellows. The primitive forge might have a flue or might not. It might have an iron firebox cast back east or one of his own forging, riveted together. Or he might have used a grate of iron bars set into the stonework, and no firepot at all.

Tongs would perhaps have been brought west, but most likely space in the wagon was reserved for tools that were harder to make. Taps and dies for threading handmade bolts were rare and prized, and a smith was well equipped if he possessed these.

Iron was in short supply, and often steel was available only by close trading or by the slow process of case-hardening in the forge. The frontier smith traded habitually, for food and supplies, for worn wagon tires and pieces of horseshoes to be shaped into the needs of the new community. Charcoal was his fuel, burned in sealed kilns, unless he was in coal country.

Crafting Everyday Items

The pioneer blacksmith made the shoes for the settlers' horses, and the chain for their harnesses, and mended the broken iron braces for their wagons. He hammered a new flintlock for a long rifle, or straightened its barrel, bent perhaps in a border brawl. The smithy became a stop along the lengthening wilderness road west, where the ravages of travel were patched and mended and the iron made whole again.

It's hard to imagine how little the smith had to work with. In a situation where a simple bar of iron was almost nonexistent, he might hammer half horseshoes, worn in two at the front, into nail rods for nails for a split shake roof. Or he might shape a salvaged brace from a splintered wagon that could go no farther into an auger, case-hardening it into crude steel. And this same brace could have been part of a wagon tire before, or a worn axe, and something else before that.

The smith made it all, including household fixtures like swinging fireplace cranes, skillets, pots, forks, spoons, and knives, as well as the files to sharpen the knives. He made saws for the carpenter, hammered and hardened; mill picks to dress the miller's stones; and iron collars for the wooden shafts in the mill.

An early country black-
smith's work, this
double hook with
upset ornamental
ends was on a
New Hampshire
farm wagon.

Forged scale, probably
used in some business
situation, which
includes the long
marked balance rod,
beautiful hooks, a
decorative spade at
the end of the rod, the
brass weight, very
good chain work, and
interesting means of
attaching short rods to
alter the scale markings.

This nicely primitive chain hook strap is one of two on an Ozark mountain farm sled. The shape, hammered from an old wagon tire, kept the chain from slipping out when slack.

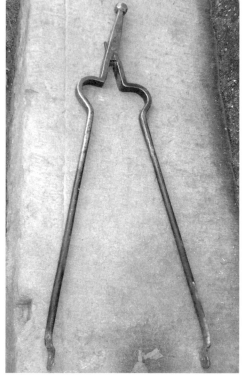

Beautifully forged fireplace tongs which work smoothly as well as being artful. The workmanship at the hinge is especially well done.

A really well-forged pair of scoops, most likely used in something like rendering, or in a tavern kitchen. The straining spoon's holes are punched. The bowls are hammered to shape.

Of Iron and Steel

When iron became more common, the blacksmith poured the bearing metal for the sawmill shafts and hammered buggy axles and candleholders. As plows evolved to steel from the old wooden ones, as the circular saw came into being, and as the steam engine came along, the tasks of the smith changed. As the age of machines grew, he became a specialist, building and repairing them, until those same machines finally forced the smithy to close, ushering in a century of stamped-out, cheapened iron.

The frontier smith might trade for crude steel of a doubtful carbon content from back east, or he might venture into foundry work to produce his own steel if he had the knowledge and equipment. Most likely he would place a bar of wrought iron in a case of iron or clay, packed with ground-up bone or old dry leather or even hair. This case went into the forge along with the pieces he was shaping, and stayed red hot for many hours. Gradually, the red bar of iron inside absorbed carbon from the organic material around it as the material charred. The surface of the bar became steel, deepening as the treatment was repeated. When the bar was forged, the hammering spread the carbon steel throughout the iron bar to raise the overall content. This was case-hardening, a make-do way to produce a very limited amount of steel.

Sometimes the knife or axe or chisel was forged first, then case-hardened so that only the outer layer was steel. As the tool later wore and

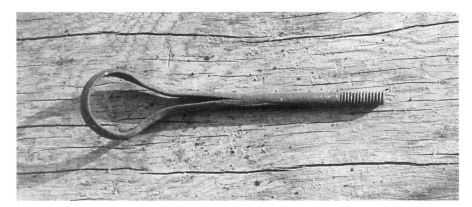

A neat bit of early forgework, part of a bucksaw turnbuckle. It was made by hammering the loop of half-round iron, then forge welding the ends, rounding, and threading the left-handed threads for this half of the turnbuckle.

A hand-forged log-skidding hook from the mountains of Kentucky. The point was driven into the log, with a chain attached to the ring. Two were used together, as a sort of free-form pair of grabhooks. This was a gift from Bill Chatfield, who I helped restore a log house, and for whom I forged two sets of andirons.

was sharpened, the steel gave way to soft iron again. In the backcountry today, there are still a lot of old-timers who believe that you will eventually wear or sharpen a modern axe or knife down past the good steel to soft metal underneath.

The Mysteries of Metal

The old blacksmith was often accorded an esteem far out of proportion to his knowledge. The term "black metal" refers only to iron, as opposed to tin or silver or pewter — white metals. But the early pioneers always believed that there were mysterious steels that could be contorted and return to their shapes, knives so sharp they'd cut into wood when laid on it, tempering fluids that imparted properties to the steel that defied wear.

Of course, the old smiths did nothing to discourage these beliefs. Some tempered in ox blood, some in mercury, for the supposed superior qualities they gave the steel. There's a story of one illiterate smith who bought at great price a recipe for a quench guaranteed not to crack the steel and to make it harder than any other. He had the written down formula, and had it filled by a chemist, who was evidently in on the swindle. After many years of successful but costly use of the magic liquid, the smith revealed the formula to a trusted associate who could read. It called for so many parts aqua and so many of chloride of sodium. That story's been around among smiths for a long time.

Today we know that the liquid for quenching steel has only one function: to cool steel of a given carbon content faster or slower. Three basic liquids are used. Most smiths agree that plain water is fastest, for a harder steel, but its fast cooling can crack steel. Salt water, that mysterious formula, slows cooling somewhat, for a tougher steel that is not as hard. Oil tempering is the use of an oil quench that cools more slowly, for less hardness in the steel and more toughness. It all has to do with the boiling points of the liquids, nothing else.

On the other hand, you can read in half the books on the subject today that salt water gives the hardest quench, plain water is next, and oil is for toughness. My advice invariably is to try it yourself when you get to hardening and tempering steels. I will confess that I never use salt water at all, getting exactly the results I want, even in delicate toolmaking and knife making, from plain water or oil quenching. But lots of modern smiths quench only in salt water, and their method works for them.

The few early smiths who really knew this weren't about to admit it publicly. Their reputations rested on their supposed knowledge of the mysteries of the iron, and this set them apart from the ordinary farmer or laborer. If more smiths had indeed known what was really going on in their often hit-or-miss craft, it would all still have been a grand ongoing joke.

Making Do

The amazing thing is that early smiths were able to do as well as they did to keep the settlers supplied. In fact, folks were just used to living with iron that wore out soon, with steel with brittle places in it that broke, and with workmanship that, while not perfect, was a vast improvement over their own abilities. The demands of frontier life were simple, and the smith's work seldom had to be of machine-shop caliber to meet those demands.

Most farms and plantations had their own shops, and took care of most of their own needs. The commercial smith was usually located in a settlement, where other folks tended to specialize and didn't have the time, inclination, or ability to do their own iron work. The farmer-smith shod his horses, made his chain and hinges, sharpened his plows, and steeled his axes as far as he was able, turning to the specialist in the village for the jobs he couldn't handle.

Many an old homestead in the country today has a small forge back among the junk in the barn, and maybe an anvil half covered with hay. Grandpa did his own blacksmithing, and a lot for the neighbors, but Papa went off to town, and nobody's touched that stuff for years.

Some years ago in Missouri, my neighbor Terry Burke was leveling the stony hillside in the woods outside his new house when he struck iron. It turned out to be a buried anvil, and more digging uncovered several sets of tongs, a froe, and some horseshoes. He called me to find out if I knew of a blacksmith shop there, because he'd found no trace of a building.

Well, this is only a couple hundred yards off the old stagecoach road around the mountain, at a spring where the little branch of a creek gets going. Most of the tools suggested a horseshoeing operation, so the earlier settler probably did some business with travelers on the stage road. There were traces of a bridge across the branch, and faint traces of a road up out of the hollow. The smithy probably collapsed with age or burned down many years ago. I traded a hand-forged broadaxe and some other homesteading tools for the forged Peter Wright 113-pound anvil. Terry split shakes with the froe, just as his predecessor on that site did.

This hollow, a tangle of second-growth woods when we came in 1972, apparently held a dozen or more cabins, and the old smith probably served them all, along with the transient trade. A washed-out mill dam farther down creek suggests another early craftsman. And the old log church still stands, down where our branch meets a larger creek.

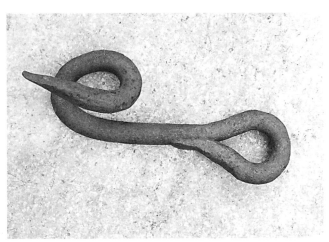

An early forged rope hook, probably shaped this way to attach to a taut line, keeping it secure, as it slid along it.

Modern-Day Pioneers

Your own experience setting up a place in the country will parallel that of other early settlers, in many respects. So will that of establishing your smithy, whether exclusively for your own farmstead use, to supply your rural neighbors as well, or even for commercial projects. You probably won't want all primitive hand tools, like the living history museums try to use, but if you're out past electricity, this is the course you'll have to follow, more or less.

I use a combination — a hand drill press and a hand blower on the forge, but an electrically powered trip-hammer, and sometimes an acetylene cutting torch and an arc welder. I make my own tongs and chisels and punches and lots of other tools, but I buy a hammer head at a junk shop, and rarely pass up any cheap piece of blacksmithing equipment, even though I could make it myself.

But whether or not you are a modern-day pioneer, whether or not you use powered equipment, whether you are located in the woods or the suburbs, yours is a kinship with the frontier blacksmith. The hammer striking the forge iron, hot on the anvil, has never changed.

Wrought-iron singletree with decorative snake hooks, from New England. The hook shape kept the trace chains from falling out.

2

GETTING STARTED

What can you expect from blacksmithing today? What will you be able to do with this demanding craft? What does it involve?

Good questions, all. Put simply, the craft of blacksmithing allows you to make by hand anything of iron or steel, depending on your skill level. Whether your interest is in antique tools, ornamental iron, knife making, forging useful household or workshop items, or even gunsmithing, you will be able to satisfy that interest. As you build your skills, you'll tackle more complex forging procedures, for pieces of greater usefulness and/or beauty.

To smith, you'll need a place to work, some tools, and a few materials, much like a beginner's woodworking shop. But unlike that setup, which often grows into acquiring advanced and expensive machinery, the smithy can stay pretty basic. You'll need a forge and blower of some type, to heat iron and steel to about 3,000°F. That's hotter than a fireplace or stove. You'll need tongs to handle that hot metal. You'll need an anvil to hammer it against, hammers to do that with, and maybe a vise to hold the iron while you bend or twist it. A handful of other small tools will set you up nicely.

If you should succumb to the equipment/tool mania that seems to go along with other crafts and hobbies, you'll find that you can make most of these yourself. And, perhaps best of all, blacksmithing tools don't need to be upgraded constantly. Hammers do not become obsolete, and anvils haven't changed much in a thousand years or so. Blowing on fire to make it hotter is an old trick, too.

An efficient shop is the blacksmith's most useful tool. You can operate under a tree or in an open shed, but a covered space devoted to your craft makes the work flow smoother. I built this shop in Hollister, Missouri, in the 1970s. Here, I made and repaired tools for myself and others, as well as conducted workshops for both the amateur and the professional.

If you're a totally untutored beginner, this craft will probably look complicated, even intimidating. All those secrets it took a lifetime to learn. Those judgment calls. Those years of practice. Maybe you should take up birdwatching instead?

It's not that difficult. It is easier if you like to work with your hands, if you enjoy making things. It's a slow business, this shaping iron. But that's part of its appeal — the investment in sweat and aching muscles that few others will make. And, of course, the pride in achievement that comes from using simple, even crude tools to produce pieces never quite to be duplicated, anywhere.

Start small. You may even stay small, forging just a few hooks or trivets or maybe a pot rack or fireplace poker. Nobody said you had to try to become a master smith. Enjoy what you're able to do well.

Try to learn from another blacksmith. I experienced a real breakthrough when I worked with the late Shad Heller 30 years ago in Branson, Missouri, after years of dabbling. Reading a few books helped, too, but there's no substitute for getting your hands in it.

So let's get to it. If you find you don't really like blacksmithing after all, there *are* all those birds out there that need watching.

Gathering Your Tools

Setting up your forge and acquiring the tools you'll need is an exercise in invention. What you can't locate at auctions, garage sales, or in junkyards, you can make yourself — and the more complete your shop, the easier it becomes to make more tools.

Setting Up Your Forge

The forge is, of course, the key to the shop. It has to be simply a contained fireplace with an air-blast channel, called a tuyere (*too-yair*) up through it. It can have an ash dump, a flue to carry off smoke, and a clinker-breaking device (a clinker is a chunk of metallic slag melted out of the coal). The forge can be iron or masonry or both. It can have a bellows or blower that's powered by hand or motor. New forges are available, but they are prohibitively priced, and you should learn early to fabricate your own equipment. Let's look at some alternatives.

I use three forges. One is a simple pan-type farrier's forge with just a heavy screen at the bottom through which the air comes up, an ash dump, and a small sheet-iron hand blower, all set on pipe legs. It's light, and handy for transporting to craft fairs and workshops. I have worked steel into a nine-pound wood-splitting maul with it.

The second forge is of my own construction, using a small but sturdy cast-iron blower, a cast firepot, and a tuyere with a triangular clinker breaker and an ash dump. The fireplace is steel plate, curved, with low sides and a cutout for long pieces. It's on three legs of recycled reinforcing rod. This forge is heavier than the pan forge, and can handle bigger work.

The third forge has a heavy, deep cast-iron firepot set in stonework with the slitted round tuyere/clinker breaker and ash dump. It's charged by a separate Champion 400 hand blower on its own stand, bolted to a concrete slab. This forge is four feet square, with a reinforced concrete fireplace cutout for long work. We heated a 70-pound anvil on it to harden it on one occasion, so it can handle heavy work. Its stone side forms part of the low stone wall of one side of my shop.

My brother John has a small forge made from an automotive brake drum, charged with a small sheet-iron hand blower. He does all the forge-work for his farm, and a limited farrier business on it. His blower, attached to a pipe with an ash dump, came from a junkyard for $10.

Making a Simple Forge

I'd advise starting small, with a forge of your own devising, to avoid spending much money. You will soon find your own particular needs, and can be on the lookout for better equipment, or make it yourself. A large brake drum has a good shape for a small pan forge. Braze legs onto its outer rim; it's cast iron, so brazing is the safest way to go (brazing is welding with brass, so the cast won't crack as it cools). If welded with arc or acetylene, the cast iron should be preheated, then allowed to cool slowly afterward. Also, slant the legs of this forge outward to keep it from tipping, since it is so light in weight. Next, bolt a disc of steel plate over the big hub hole in the bottom, through the wheel lug-bolt holes. Bore G- to H-inch holes in a circular pattern in the center of this plate to let the air come through and for the ash grate.

Next, weld or braze a 2-inch water-pipe tee to the plate disc, and thread a 12-inch piece of pipe from it for the blower. Then you can use a pipe cap or collar and plug at the bottom to seal it. It is easily removable to dump ashes. Weld a bar across the plug to make it easier to unscrew.

Blowers

Shop around for a small hand blower. They can often be found rusted up, or with the handles broken, very cheap. The fan will usually be sheet iron and some of the blades might be bent, but you can straighten them. It's harder to weld or braze replacement blades on and keep the fan in balance. You can soak the bearings loose with penetrating oil.

I recommend a hand blower because you have complete control over the fire. An electric blower will free your other hand, true enough, but you will burn lots more coal with an electric blower, and a moment's inattention will let you burn carbon steel or thin mild steel, which you'd have had your eye on if you were hand cranking.

A leather bellows is the oldest form of air blast we know of. Aldren Watson goes into great detail on the construction of the bellows in his book *The Village Blacksmith*. The bellows takes a lot more room than the gear blower, but is more nostalgic. It was most often used where the blower wasn't available, as when a pioneer built his charcoal forge out in the wilderness. The double bellows does keep on blowing for some time after you let go, however, which frees your hands for tricky things like forge welding.

If you brace your blower pipe to the lip of the forge pan, you can mount the blower directly on it. My pan forge is done this way, with a bolt through the pipe. The midsize forge blower is mounted on a plate. Also, if there is no clinker breaker, you must take the fire apart every time to clean it before you forge weld, since there must be no clinkers for this operation.

Otherwise, this lightweight forge will get you started and allow you to do most jobs around your farmstead. Keep the forge clean and out of the weather. Rain on burned coal will cause sulfuric acid to seep down from the sulfur and eat out the bottom of your forge very soon.

Building a Simple Forge

1. A simple forge for most jobs. First, two pairs of legs of recycled reinforcing rod are welded together.

2. The legs are welded to a piece of flat steel plate about 8 inches square.

3. A piece of 2-inch water pipe is welded to the bottom of the plate, and a hole cut in the side, where a piece of 1¾-inch conduit is welded.

4. An automotive brake drum is centered over the plate, and holes are bored to bolt through; orit can be welded in place. A grid of holes is bored to open into the pipe up from below. This is a top view.

5. The forge nearly complete. The broken handle of a junked blower was welded and will get a wooden grip. The ash dump can be a threaded pipe cap or a weighted hinged trapdoor. A light steel rod brace strengthens the blower mount; it's bolted to the blower mount and welded to the cast-iron brake drum with bronze or nickel rod. Care should be used not to let the fire burn shallow, or the plate base of the forge will burn or melt. This forge took less than four hours to build; materials cost less than $5.

Masonry forge firepot, showing a tuyere/clinker breaker inside, with the handle weighted for proper air blast, air inlet, and ash dump at right. Cutouts at left (really the top, in use) allow stock to go lower in the fire.

A Champion 400 blower on a stand, which I use with the big firepot in my masonry forge.

Handmade, 4-foot bellows, belonging to Dr. David Morris, of Earlysville, Virginia. This is a double-action bellows, with the center panel held stationary and mounted horizontally. The missing handle pulls the bottom panel up to force air out of the nozzle and up into the top chamber, where a weight pushes it out after the handle is released, freeing both the smith's hands.

Anvils

Next, find an anvil, and few tools are so hard to find. Like the forge, which can be bought new, the anvil is available, but it can be costly ($2 to $4 per pound, in 2005 dollars). There are two kinds of the London pattern anvil we're most familiar with. One is the cast-steel anvil, which is one piece, with the top hardened by chilling as it is cast. Cast anvils tend to ring nicely, but that means next to nothing in determining their efficiency. (A brittle cast bell rings even better, but you can't hammer iron on it.) The only part of the anvil that should be hard is the top surface, or face. When inspecting a used anvil, look for chips or cracks in the top, which I find to be most common in cast anvils. We'll talk about repairing them later.

The other kind of anvil you're likely to see is the forged one. Forged anvils were formed in three pieces: the base, body, and top plate. These were welded together sometimes as a technical school project, with strikers (blacksmith's helpers) hammering the heavy sections at welding heat. The top plate was of steel, the lower parts of iron. The anvil was then chilled and tempered, affecting only the face, to harden it.

These anvils are rarer today than cast ones, and they represent a high achievement of the metalworker's craft. I find that these anvils seldom ring sharply, being mostly soft iron, but are quite as good to work with as the best cast anvils. Common faults are chipped or broken face, or even a separation of the weld of part of the plate.

You will hear much about the ringing of the anvil. The best one I ever used rang not at all, but it had a perfect face that was hard enough to bounce the hammer nicely. It was a forged anvil, on permanent loan from my friend Bill Cameron, and I recall it fondly. But it rang not once.

The overall hardness of a piece of steel or cast iron causes it to ring when struck. Cast bells ring nicely. So does brittle cast steel, high in carbon. And while it is true that an anvil that rings nicely is good to listen to, it just might break under a heavy sledgehammer blow someday. Don't make the sound your only criterion.

Any anvil will wear with use, so you may well find a specimen with a wavy face, or one sloped off sharply from drawing and scarfing (procedures we'll explain later), the smith having used the same side over and over. I work from every side of the anvil, so I can often use a worn anvil, concentrating my work on the good surface that is left.

Refacing an anvil is always possible, though the old surface may have to be built up or milled down to be flat enough to weld a new plate onto. If it's to be milled or ground flat, the face should be annealed, which means heating the anvil red hot. A big brush fire with logs and stumps in it will generate enough heat, then the anvil should cool very slowly. Afterward you can grind it flat with a portable "sidewinder" body-shop grinder, or have it milled.

If there's enough plate left on a forged anvil, or if yours is a cast one, the plate can be rehardened after milling. It means red heat again, at least at the face, then quenching in a very large tub or creek, then tempering the top to the desired hardness, usually a purple or bronze, with a torch or by

This is a forged London anvil, showing clearly the joint at the welded hard plate, or face, on top. It's also welded at the narrow waist. At left is the horn; at right the tail, with the square hardy hole and smaller round pritchel (punch) hole. The step at left center is not hardened, being used for cutting, with hammer and chisel.

suspending it upside down over the forge. Alexander Weygers tells how a small homemade anvil can be tempered in his book *The Modern Blacksmith*.

Tricky business. A new plate of carbon steel can be arc welded over the anvil surface, and not hardened at all. Eighty-point carbon, which means it contains 0.8 percent carbon (high carbon), will give you a fair surface to work on and usually won't chip, though it will eventually dent.

I have a 300-pound cast anvil that my friend Bill Cameron found for me at three cents a pound in a Joplin, Missouri, junkyard. It was broken across the corner through the pritchel hole, sloped clear down through the hardened top surface on one side, with a chunk broken out of the top of the horn. I heated it and built up the rounded top corners with an arc-welding rod, using some fifty ³⁄₁₆-inch rods. Then I welded a piece of steel into the horn and built up around it with rod. Finally, I ground everything flat and smooth with a sidewinder grinder. I'm on the lookout for a five-inch plate heavy enough to weld on top, through which I'll cut the new hardy and pritchel holes.

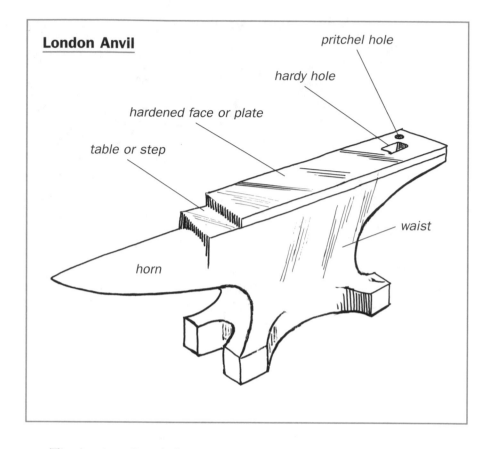

London Anvil

pritchel hole

hardy hole

hardened face or plate

table or step

waist

horn

That's a lot of work for an anvil, but it's still cheap. And I got a lot of use from it before I started the restoration.

It's best, of course, to find a good anvil and pay the extra money for it. A hundred dollars will usually buy a fair one at auction, or from the rare farmer who'll sell you the old one out in the barn. And considering that you won't wear it out in your lifetime, it's a good investment.

Or you may use a section of railroad rail, cut to a point with an acetylene torch and ground smooth. This is nice steel, and will do quite well for small jobs. I've made several of these, sometimes welding a plate on top, with the hardy and pritchel holes in it, to overhang the tail of the rail anvil.

You'll find the shape of the London anvil, the shape we see most often, handy for your work, which will soon let you see how inconvenient a simple block of iron or a square-cut section of rail would be. Even a worn or chipped anvil will let you do some quite good work. I'll usually use a worn anvil a ridiculously long time before I undertake the replating or milling necessary to put it back in shape.

If you're the type who makes friends with the neighbors easily, you may be able to get on good enough terms with an old farmer up the road to borrow his anvil, particularly if he's not using it but has too much attachment to sell it. It's hard to ruin an anvil, and if you exercise care you won't even dent it with long use. And if you sweeten the deal with free work for the farmer now and then, a loan like this can be parlayed into a sale, eventually, or even a gift.

It is a shame so many anvils were gathered up and sold as junk iron during the fevered days of World War II. Our parents or grandparents often patriotically gave away equipment they could not foresee as ever being needed or worth anything again.

The anvil's mass is as important as its shape, so if you fabricate one, make it heavy. A railroad rail anvil is better if you weld slabs of steel on each side of it for more mass. I have shaped small anvils from odd pieces of iron of all kinds, using a forge, a cutting torch, and an arc welder. Whatever you do, the anvil is one of the two pieces of equipment you'll use most in your shop, so keep at it till you have a good one.

A Variety of Vises

The vise is the next heavy tool you'll need, and I've never seen an improvement on the old leg, stump, or box blacksmith's vise. The long leg lets you stabilize it better than any flat-bench mounting for a machinist's vise, and it's meant for light hammering on, as well as holding for bending and twisting.

A 6-inch blacksmith's leg vise. This one was a gift from Bill Cameron. It was missing the spring and mounting bracket, which I forged.

I have four-inch, five-inch, and six-inch leg vises, only one of which came complete. When searching among the junk for one, see that the screw is good and tight, not worn badly, and that the jaws are in fair shape. Everything else is easily repaired. There's a method given in M. T. Richardson's book *Practical Blacksmithing,* a collection of early procedures from individual smiths, for rethreading the screw, but it's not easy.

Most often the opening spring will be gone, as will the mounting bracket that also holds the spring. The bar handle that is intact will also be rare. I replaced a bar on a family heirloom six-inch vise, complete with ball at each end, for my friend Dudley Murphy some time ago. The vise belonged to his wife's grandfather, or I'd have tried to talk them out of it.

If the jaws are loose so that the movable one flops from side to side like a worn pocketknife blade, all you need to do is tighten the joint. A big washer will sometimes do, but you may want to heat the joint in the forge and hammer it closed. Remember that the vise is carbon steel, and don't quench it to cool it after heating or it'll be brittle.

The jaws themselves will be worn to some extent, and if badly, you can arc weld bars of medium-carbon steel over them. You can also groove the faces of the jaws for better holding, but this will put a pattern in the hot iron you shape in them.

The mounting bracket is a simple piece, with a slot for a tapered wedge to hold the vise tight. Since the fixed jaw upright that the bracket holds is tapered itself, you can shape your bracket smaller, then drive the upright down into it. The return spring goes into this bracket as it comes around the upright, and extends down, to push open the jaws from just above the pivot.

I use leaf spring for the vise return spring, hammered to shape, then quenched and tempered to a blue. See chapter 5, on tempering, for specifics. Hammer ears on the spring to keep it in place, or it can slip past the movable vise leg and do you permanent damage as you stand in front of it.

Leg vises are going up in price now, being more and more antique-shop items. I've bought leg vises for as little as $10 recently and seen them as high as $85.

A machinist's vise is a good enough substitute but will cost more, and it doesn't allow the freedom of space for difficult bends, being mounted on a bench. The T-shaped head and jaws of the leg vise let you bend hot steel down into a tight U shape, or form a hoop, or do most other blacksmithing jobs much easier.

You may forge any of several shapes of jaw surfaces for your vise; such surfaces can be slipped over the existing jaws for holding odd shapes. These are handy for ornamental ironwork, where you're doing compound bends and twists.

Hammer Choices

You'll need hammers of all sorts, but I'll list a few that will get you started. Depending on your brawn, your mainstay will be a straight or cross-peen hammer of two to four pounds. I use a three-pound hand-forged hammer. But forging hammers is more work than it's worth. Flea markets will yield usable hammer heads for a dollar or so each, and you'll spend most of half a day forging one.

Add a smaller hammer, about 16 ounces for close work, and a 6- to 10-pound sledge for when you have help. A small ball-peen (12 ounces) will let you rivet and decorate, and a 2-pound one will get bigger jobs of this kind done quicker.

Indispensable Tongs

Tongs are limitless in design, as they are developed for specific kinds of work. Quite often, you'll find special tongs no one can tell you the original use of. Some inventive smith probably made them for one of his specialties, which may have been a one-time project or a way of working that died with him.

For close work, like knife blades, I use a tightly closing pair of short-handled, 10-inch tongs my brother made for me from old horse hoof nippers. A longer 18-inch pair is for bigger work, and its flat jaws barely close completely. Another pair the same size has rounded jaws for holding round stock. Yet another 18-inch pair has diamond-shaped jaws for square stock.

A very handy pair has narrow jaws with a rounded notch for holding chain or other small stock to be forge welded. A flat-jawed pair will let the white-hot iron slip before you can hammer it together. Another special pair has one curved, grooved jaw and a short, angle-bent notched one for holding the chain link while you shape it; the pair was designed by my friend, blacksmith Shad Heller. Shaping chain links with flat-jawed tongs is a pain.

A collection of smithing hammers. 1. 1½-lb. ball peen. **2.** very light cross-peen. **3.** 2-lb. ball peen. **4.** 3-lb. welding hammer. **5.** 3-lb. 45-degree peen (right-handed). **6.** 8-lb. sledge. **7.** 1-lb. ball peen. **8.** 1-lb. cross peen. **9.** 6-lb. sledge. **10.** 12-oz. ball peen. Hammer heads are put on hot, with the holes expanded, then wedged.

I have dozens of other pairs of tongs, some quite large, that I've collected over the years. One pair is notable — it's a wide-jaw set, with right-angle bends inward at the jaw tips, for holding hammer heads or axe heads at the eye for forging. Quite handy.

You'll find special wagon-tire tongs and other pairs that won't do you much good in general work. You can forge any of these yourself from mild steel or medium-carbon steel, unless you have a good cheap source for the tongs. I figure that a pair in good shape that costs under $15 is a good buy, because it takes me well over an hour to make a pair. Of course, the pair I make will be more tailored to my specific needs, so more valuable to me. My beginning students usually take over half a day to forge a pair; this is about right.

Forging Your Own Tools

Hot cuts, cold chisels, hardies, punches, and drifts all can be forged easily. The hot cut, which is a chisel with a handle, is hardest, because punching the hole for the handle will take time. An old hammer head is a good piece of steel for this, with the striking face softened so you won't chip off pieces of jagged steel.

Hammer or axe-eye tongs. These are handy for holding the head of the tool, through the handle hole. I use them a lot in forging splitting mauls from old sledgehammer heads.

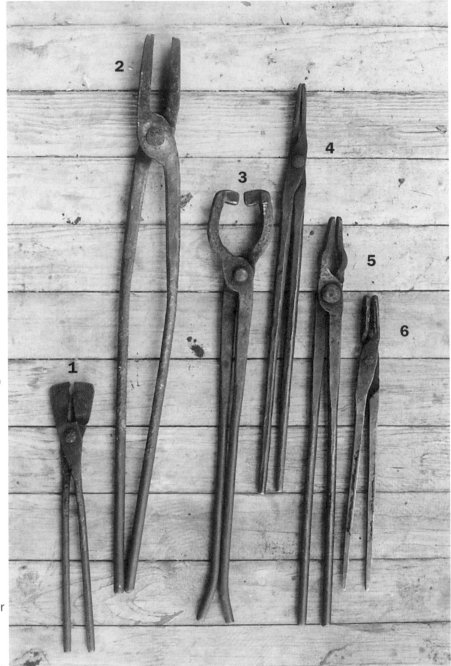

The tongs I use most. 1. chain link welding tongs. **2.** large, lengthwise grooved tongs. **3.** hammer-eye tongs. **4.** flat-jawed tongs. **5.** round stock or bolt tongs. **6.** short pair with V-groove for small general work. All these are hand-forged.

I'm reclaiming a pair of rusted tongs here, while my longtime helper, Bill Cameron, works the portable forge. I've freed the pin and am bending the heated jaws for a better fit. Good used tongs are worth buying, since you'll spend half a day or more making your first pair.

The hardy, an anvil-mounted chisel, is easily forged of automotive axle steel, which is medium carbon and quite tough. You don't want this tool hard, because sooner or later you'll miss and take a chunk off it with the hammer.

I recall forging a 12-pound digging bar for Bill Cameron one day. We were cutting 1¼-inch square stock, with him holding the iron on the hardy and me swinging the short-handled 6-pound sledge. I hit short, glanced off the corner of the hardy, and the hammer flew out of my hands, smashing into my right leg just below the knee. Bill said I chewed up a lot of grass around the shop before I calmed down. If the hardy had been brittle, I'd probably have had shrapnel to deal with, too.

With axle steel of 30- to 50-point carbon, tempering isn't necessary. Just hammer the hardy to shape, fitting it both ways into the hardy hole (which won't be exactly square), then anneal or normalize to sharpen. We'll talk

Punches and chisels. 1. ⅜-inch round punch. **2.** standard cold chisel. **3** center punch. **4.** ⅛-inch punch. **5.** square punch. **6.** round drift. **7.** long-taper cold chisel for soft metals. **8.** awl for punching sheet metal.

more about this process in chapter 5. Heat to cherry red and quench. There isn't enough carbon in the steel to harden it to a brittle stage. If in doubt, do draw the temper to a bronze. You can always reharden if it's too soft.

If you want a cold hardy, one for cutting steel cold (like a cold chisel), it must be harder. Use at least 80-point carbon, as in heavy leaf or coil spring steel. Or jackhammer drill steel, which is higher in carbon. Quench and draw the temper to a purple. If that's too soft, heat, quench, and draw to bronze. It's better to have a tool too soft than to have it break. Again, I'll explain all this more in chapter 5.

Set tools (those struck with a hammer) and others. 1. square drift. **2.** round drift. **3.** handled cold chisel. **4.** hardy. **5.** file. **6.** small flatter. **7.** fuller set (the right-hand piece goes into the hardy hole, where the hot metal is laid on it and struck with the other piece to make a stepped-down, even indentation). **8.** hot chisel. **9.** nail and rivet heading bar.

Miscellaneous tools. 1. hacksaw. **2.** steel tape. **3.** adjustable square.
4. tin snips. **5.** bullet-mold ladle used as a forge dipper. **6.** forge poker.
7. coal shovel.

The Trip-Hammer

If you do a lot of forgework (if, say, you plan to make your living at it), you'll soon wish for a trip-hammer. This is a ponderous mechanical hammer that strikes a series of very rapid, heavy blows on its own anvil. It's foot-operated, so it lets you have both hands to hold things. It's a godsend in forge welding separate pieces, and it will speed production immensely.

I have an ancient one that originally was powered from an overhead line shaft with flat-belt pulleys — probably steam powered or even water powered. Its last owner was metal sculptor Henry Menke, who'd owned an ornamental iron business. He powered the trip-hammer with a three-phase electric motor, which I replaced with a single-phase two-horsepower model. I have about $200 in it to date, and it's worth every penny.

Bob Patrick, then of Bethel, Missouri, bought a later-model trip-hammer at auction for $50, mostly because no on else knew what it was. They're always cumbersome and impossibly hard to move, so you can sometimes get one cheap if it's in the way. They shake any building when they're running. Also, you can get maimed in one.

The Grinder

You can keep acquiring blacksmithing tools and equipment till you have a warehouseful. But add a grinder of some sort to the equipment we've talked about and you're set for most jobs, even without a trip-hammer. A grinder can be a slick, shielded, double-shaft electric one or an old hand grindstone.

A cheap, powered one is simply an electric motor with an arbor adapter and an abrasive wheel. You can even find a motor free, usually, if the starting mechanism is bad. There's no trick to giving it a spin as

David Morris's hand drill press, almost identical to mine. He's modernized it with an added universal chuck. The ratchet feed at top is adjustable; it keeps the bit biting the iron.

My trip-hammer, not set up for use. The bail, or foot control, partly visible at the bottom allows the smith to use both hands to hold the pieces to be worked. This is an ancient, 1½-horsepower model by Novelty Iron Works, in Iowa.

you turn it on to start. My lawyer friend Turner White had a manual starter in his blacksmith shop and so do I. Get at least ½ horsepower or you'll bog it down.

Beware of secondhand grinding wheels, particularly if you plan to turn them fast. The standard 1,725 revolutions per minute will keep most stones intact, but a geared-up speed can shatter one and kill you, as one did a machinist acquaintance of mine. And have a toolbar near the wheel to rest the work on so you won't jam pieces into it.

A handy tool, this shear is made of two lengths of cutting-edge steel, with a square shank for the anvil hardy hole. Iron or steel is cut while red hot, with downward pressure and hammer blows if necessary.

Other Tools You Might Want

A shear will let you cut hot iron and steel straight and smooth. It's like a large pair of tin snips with long blades, to be mounted conveniently in the hardy hole of the anvil. Instead of handles, the top blade is operated by hammering down along the stationary bottom one for the cutting action. Or the top blade can have an extended handle from its tip for hand leverage in cutting.

Another unnecessary but handy tool I have is an antique drill press. It has a big snake-spoke flywheel and hand crank, with an automatic ratchet feed. You'll punch most holes hot in your forgework, but there are times when this is awkward, and the drill press is useful. Simpler variations can be had for $50 to $75 at farm sales.

A tap and die set for coarse and fine threads will let you make bolts and nuts for unusual uses a lot cheaper than at hardware-store prices. With the swing to metric sizes, more old sets are becoming available. Try the tools out to make sure the threads are not worn beyond use. One of the handiest uses of a tap and die set is as a thread chaser, to clean the threads of rusted or battered bolts or nuts.

A bit of advice here. Learn to use these hardened-steel tools before you break them or wear them out from ignorance. A small tap will snap if forced through a hole dry or too quickly. Alternate forward and backward twists with plenty of cutting oil to keep the tools cool, sharp, and intact. Visit a machine shop and watch the techniques.

Building the Smithy

Finally, you'll want to house your workshop effectively, but cheaply. I built one in the woods at our place, a 12-by-16-foot shed-roofed, open-sided version of the pole barn. It will later be enclosed for winter work, with double doors for getting bigger jobs out of the rain. It began with just four timber uprights, set on flat stones on the ground. A post hole next to each timber was filled with concrete, into which went a heavy metal bar, bolted to the upright, with a cross pin down in the concrete.

That keeps the structure stable, without letting the timbers rot (as they would set into either the ground or the concrete; water would run down into the crack between the concrete and wood and rot the wood). Overhead, these uprights are notched and pinned to sloping stringers (end timbers overhead) that overhang front and back. Purlins, which function as rafters, are pinned to these, and galvanized roofing is laid on the purlins. The whole thing is angle braced.

The shop took two days to build, but that included hewing the timbers and working the notched joints. It could have been spiked up in a day using round poles. The roof cost $70, which could have been nearer $50 if I'd been able to get in on a special sale of corrugated roofing. The other materials were growing on the place or in my scrap pile. One wheelbarrow load of mixed concrete filled the four holes, which I widened at the bottom to give more stability.

The commercial shop I built in Hollister, Missouri, was also small — 12 by 16 feet again — but was framed with 8-by-8-inch timbers, with a peaked, shingled roof. It had a 3-foot stone wall around it, with an opening for a wrought-iron gate. Part of the wall was the masonry forge, which was built in. This shop was more elaborate, and was built to allow visitors to watch the work inside. It also conformed to the stone-and-half-timber building style of the English-looking town.

In spite of this smithy's small size, I taught several courses there, with as many as half a dozen students working. Unless you have a lot of equipment or need space to store long lengths of steel, a small shop is adequate.

For a floor I used raked gravel, which is easier on feet than concrete. Security's no problem at home, but I locked up small tools at the village shop, and had the blower, anvil, and vise bolted down and spot welded to slow their removal.

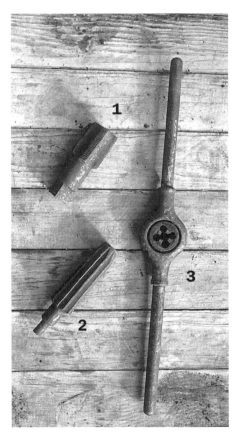

Handy odd tools. 1. threading tap. **2.** reamer. **3.** threading die. You can make your own bolts and nuts for special projects, or use them for cleaning up battered or rusty threads. The reamer is for enlarging holes cold.

Do get your operation under a roof of some kind. The spreading chestnut trees are all gone now, with the turn-of-the-century blight, and wouldn't give your smithy much protection anyway. Rainwater can get into your blower, freeze and burst it overnight, and things like files are ruined by rust.

I've heard of smiths using tepees, vans, and trailers for shops; yours, of course, is a matter of your own inclination. I know of several in the limerock country that are simply back under overhanging bluff ledges — shallow caves.

You'll develop a sense of pride in forgework well done, and this can extend to your smithy. I like wrought-iron decorative work in evidence, and construction strengthened with examples of my smithing. This is more important if you plan to make part or all of your living from your shop. Blacksmith shops are usually dirty, dark, and full of what is, to the nonsmith, junk. A relatively neat shop adorned with some of your best work will encourage commissions.

A simple touch I like is my slack tub, which is half a 60-gallon oak whiskey barrel. It's no more efficient than a galvanized tub or large bucket, but I get lots of comments on it. Similarly, I have my anvils spiked in place on their blocks with fitted forged staples, with hooks and loops for tools. Aside from the convenience, this arrangement makes people think I actually know how to use all those fascinating tools.

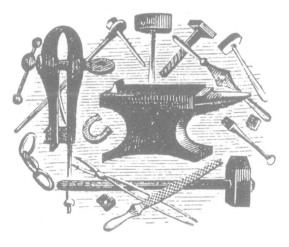

3

COAL, IRON, AND STEEL

Keeping a supply of materials will constitute most of your smithing overhead, after you've put the basic shop together. At least, you'll need lengths of round, square, and flat steel stock, along with recycled shaft steel, and possibly some old circle-saw blades. And you might want some automotive axles and springs, both leaf-type and coil, for carbon steel.

You'll need 100 pounds or so of coal, not because you'll use it up fast, but because it'll be hard to find a good source. Most of us tend to bring home anything we think we might need, be it coal or iron, and it piles up (becoming junk to the untrained and unappreciative eye).

Choosing a Fuel

There are many options for fueling your forge. Coal is the best choice, because it'll produce enough heat (3,000°F or so) for forge welding without burning up fast. Charcoal is less satisfactory, and commercial charcoal briquettes just burn away too fast. Natural or bottled gas, in a forge with a forced-air blower, will heat well, but every one I've seen makes a lot of noise. And compared with coal, gas is expensive. Oxygen/acetylene is the hottest flame we have, but that's *really* expensive when it comes to heating iron of any size. Its best applications are in cutting metal, brazing, and welding sheet metal.

Finding Good Coal

The composition of coal — anthracite (hard) or bituminous (soft) — isn't as important as its coking qualities and whether it burns clean and lasts. You see, the fire, along with sprinkled water, steams and burns out the tar and sulfur and converts the fuel to coke, which is very high in carbon, and burns clear. Only the coke should come in contact with the iron you're forging. As it burns out, you bring in more, keeping raw coal away from the center of the fire.

Blacksmith coal is simply a low-sulfur, good-coking coal, usually soft coal, in pebble-size chunks. It should last a long time and produce few, if any, clinkers. The difficulty today is that the low demand makes it unprofitable for coal companies to keep their product clean and of the high quality demanded for forgework. Just about any coal will burn in a heating furnace or power plant, the EPA permitting. (I hope the EPA can keep us breathing longer.) So why should they bother with top quality for a very small blacksmithing market?

You can use furnace coal in your forgework. It'll heat the iron hot, all right, but the sulfur fumes will lift off the top of your head, the smoke will endear you to none of even your distant rural neighbors, and you'll have a mass of clinkers in the forge very soon. Forget forge welding altogether with this coal. A clinker's crude metallic substance melted out of the coal gives off oxides, making it impossible to do good welding.

I launched a commercial shop once, in a region that had no good coal supply, using furnace coal spilled along the railroad tracks. Seems the coal for the nearby college's power plant was loaded into trucks periodically, and the drivers hot-rodded across a shallow stream as they started off. Spilled coal was almost a foot deep along the road, crushed fine by the truck tires. The coal smoked, smelled, and made my eyes smart, but I did a lot of work with it. It did burn up fast, but the price was so right.

Now, the question I'm asked most frequently by other smiths is the location of my coal source. It's been bad and it's been good.

I used a one-man coal company in Oklahoma for a while. He dug coal till he had a truckload, then came east, supplying his few but faithful shops with 100-pound sacks as he went. The proceeds went for really bad whiskey, and there was a regular cycle of booze up, sober up, dig up, and load up between supply runs. But the coal was good, until he got to loading up with surface coal around other mines to save the work of digging. I had to stop using his coal, and so did most of the other smiths in the region.

Really distant suppliers in industrial centers could and did ship good coal, but I actually paid more for the freight than for the coal. Finally,

This clinker is made up of crude metallic substances melted out of the coal. Oxides given off by it will keep you from making a forge weld, and the clinker itself will clog up the air passage (tuyere) in the forge. Clinkers appear in the bottom of the fire as dark objects.

Bill Cameron located a mining supply company in Joplin, Missouri, that had good coal cheap, and it was close enough for a trip now and then, when I lived in Missouri.

The best path to good coal is still asking other smiths. If none are around, or if they're having the same problem, try the offerings of the coal companies in the area. If this doesn't work, widen your search area. It takes only a sample bucketful of coal to see how it will burn, and trying it out is a necessity. I never buy more than a ton of coal at a time, and usually split that much several ways among other smiths. That's a fair pickup truckload.

Coke can be bought outright, but it is expensive and usually not available in small quantities.

Charcoal

Charcoal was the backwoods blacksmith's mainstay. He made his own by keeping air away from burning oak, hickory, maple, or locust logs once the fire was going well. You can do this too, or try to find log charcoal for sale. The briquettes you might use for your barbecue are pulverized charcoal that's mixed with a binder like cornstarch, a process that creates a sales-oriented shape but reduces carbon content (which should be as high as possible).

I've made charcoal simply by lifting hot coals from the fireplace and dropping them into a bucket of water. You can build a log fire in a pit, then cover it with sheet iron and plaster up the cracks with clay. The fire will burn very slowly with so little oxygen, finally going out, leaving some good charcoal, some ashes, and some wood. With practice you'll learn how big a fire to make, when to cover it, and what to expect.

Charcoal makes a very clean forge fire, good for welding. Charcoal's big shortcoming is that it burns up fast. Good coal will last many times longer. Even bad coal has more energy in it than charcoal.

Gas and Other Fuels

A gas forge is also clean, and can be built with firebrick, piped gas, and a blower. You'll have to do some advanced designing to get enough heat for welding, but this also means you can leave steel in the forge without burning up the steel. Also, the gas forge, which is sort of a contained-flame blowtorch, heats a larger area, so you can have a dozen irons in the fire at once. Good for production work. But the price of liquid petroleum gas is climbing and that's about all you'll be able to get in the backwoods. Natural gas will do well if you are on a line.

We should also mention use of the oxygen and acetylene torch to heat your metal. This is done by some smiths at crafts fairs and indoor exhibits, but it's expensive. The torch is good for temper drawing, however, which we'll discuss later.

The forge fire will burn high and smoky until it converts the raw coal to coke. Keep pulling coal up to the fire as you turn the blower, and keep a good flame to burn off the smoke. Finally, wet down the raw coal and pack it into a cave-like shape, with clean-burning coke inside. Then it's ready for the iron.

Wayne Haymes shows me the gas forge he built of firebrick. He used a squirrel-cage blower with adjustable draft, feeding air through flexible exhaust pipe. His gas pipe with cut-off valve had no orifice or nozzle, but was brazed into the side of the 2-inch steel pipe air inlet. He was able to do production smithing with 8 to 10 irons in the fire with this forge.

Iron and steel are your raw materials, besides fuel. For simplicity, I'll refer to mild steel (that which has little carbon content) as iron and medium-carbon steel and tool steel (high carbon) as steel.

Testing Your Metal

We'll talk more about choosing the right steel for the right tool in chapter 5, but you'll need to know something about the carbon content of metal before you scrounge a supply.

Carbon content affects the toughness and hardness of steel. The more carbon, the harder and more brittle the steel. To keep it simple, mild steel just isn't hard. It dents and bends and is weaker. Medium carbon is tough steel, which will bend, but not easily. High-carbon steel is relatively brittle. Let's use a theoretical scale of 0 to 100, with mild at 0 and high carbon at 100. That also happens to be the approximate point scale used to denote actual carbon content.

Just because a steel is temperable doesn't make it good for everything. I wouldn't make stone-cutting tools of 30-point axle steel, because it just won't harden enough. Nor would I use an old file of well over 100-point carbon for a knife or something that has to flex often. The blacksmith's trick of welding steel cable together to make a knife is impressive (and a neat trick), but it makes a low-medium carbon blade that is too soft for most people's taste. When buying steel, you may run into numerical classifications, such as 4140, 3130. The thing to remember is that the last two figures denote carbon content in points. This is not true of all steels, however.

The average old-time smith had little knowledge of steel's carbon content as it related to hardening and tempering, as we said in chapter 1, and he developed techniques that would have to be called crude today. Also, his supply of steel was erratic and the steel itself was not uniform, so he had another handicap. The old accounts say without reservation to temper draw an axe, for instance, to a blue, which is about right for 80-point steel. But the old handmade axe might be 50-point steel, or 100-point steel, or even wrought iron that is case hardened only on the surface. The old smiths used what they could get, and if they could get enough of the same stuff, they experimented until they could work it reliably.

You'll need to do the same, if you're using recycled metal. You can tell a lot about the steel you find if you know what it's been used for before. If the original use for the metal didn't require high-carbon steel, it was a waste for the manufacturer to supply it. So use your knowledge of mechanical stresses to guess about a piece that you haven't worked with. Working it is, of course, the best test.

I honestly don't know a quick way to test for carbon content in a strange piece of steel. Jack Andrews, in his *Edge of the Anvil,* has a brief but handy chart giving the carbon content of commonly found recycled steel items. He and several other published smiths also suggest uses for certain steels, all of which my own experience does not agree with. But I hasten to state that one smith's methods with a given steel will produce results different from those of another. And hardness preference for, say, a knife blade is almost purely a matter of individuality.

Trolling the Junkyard

By far most of my iron and steel is recycled and comes from the most commonplace (and sometimes unlikely) sources. Automotive coil and leaf springs are mainstays for high-carbon steel, and are good for knives, chisels, axes, punches, machetes, prybars, and, of course, other springs.

A coil spring of eight loops, four inches in diameter, yields more than eight feet of steel bar. A used garage-door spring gave me more than 50 feet of ⅜-inch rod, which, over the years, has become woodcarvers' knife blades, small punches, pivot pins in tongs and shears, nails, and screwdrivers. That spring cost 50 cents at the time. The heavier car springs may cost as little, if they're lying out in the junkyard. If they're attached to an identifiable vehicle, they're considered replacement parts and cost more.

For more mundane uses, carbon-steel coil spring makes a triangle dinner gong with a clearer, sharper tone than mild steel. It also makes a tougher fireplace poker, but don't quench it to cool it or it'll shatter in use. This steel also makes a dandy pinch bar or cold chisel. Alexander Weygers shows how to unwind coil springs in his *Modern Blacksmith.* I use the vise or hammer and anvil to straighten a short section of heavy spring at a time, but you can unwind light ones from a spindle when hot.

Leaf springs are so very handy, except they're often too wide. A cutting torch or a stationary shear to fit into the anvil hardy hole will cut them to size. I use them full width for froes, cutting across at the tie bolt to get stock for two from one leaf. Drawknives require only about one inch of width, so I use the shear to split the springs.

Old hayrake teeth are good steel, too. They are about the same high-carbon content as springs — 80 to 100 points. Axle steel is 30 to 50 points, or medium. Woodsaw blades are of medium-carbon content, too. Reinforcing rod production is not carefully controlled, often using recycled steel from whatever is on the yard at the time. Rebar is usually like axle steel — temperable but not high enough in carbon content for chisels, punches, or rock tools, and not really predictable enough.

You can find lightweight angle iron in old bedspring sets. Plow points are higher carbon, as is grader blade, jackhammer stem, and other steel designed for wear against stones and the earth. Old tie rods and drive shafts are about medium carbon.

Good steel is often the wrong size, though. Axle steel is a good example. One of my students wanted a bowie knife from a 1½-inch axle, and we took turns helping him hammer it. It was a little soft for a knife, but he'd broken knives throwing them, and this steel is by its nature almost a guarantee against that.

Blacksmithing iron and steel are to be had free, as junk everywhere. The long piece here was an automotive helper spring that became a froe and two drawknives. The piece under it at the other end became one of the set of andirons shown in chapter 8.

Bar stock shown here is available by the pound from any steel-yard. It usually comes in 20-foot lengths, with cutting extra. Coil springs are often free when not identifiable at the junkyard. They can be heated and straightened in the vise (below) or hammered over the anvil.

Buying Stock

I use mostly recycled metal, but sometimes I must buy stock. I keep a supply of ⅜-inch, ½-inch, and ¼-inch square and round rods, some bar stock from ¼ inch up to 1 inch thick, from 1 to 4 inches wide. Odd sizes of angle iron are also useful.

For the special order, I'll often buy an unusual dimension of iron. Digging bars are usually 1¼ inches round or square and 5 to 6 feet long. A broadaxe usually requires ¼-inch plate. And I sometimes buy new steel for making knives.

This steel is available from the steelyard, which will be located in any sizeable town. There are cutting and delivery charges, but lengths are often 20 feet, awkward unless you have a vehicle equipped for moving such lengths. A pickup truck can carry an over-the-cab rack — easily built — which is also handy for long boards, poles, and sections of pipe.

Rare metals like Damascus steel and wrought iron are hard to acquire now. Damascus always was, and its fabrication is among the highest forms of the smith's craft. You won't find it in a junkyard. Its layers of very high-carbon steel or cast iron alternate with wrought iron, welded together, doubled, welded, often 200 times or more. The legendary swords of Damascus were made this way; it was a method of getting carbon into wrought iron when this was a major problem.

Wrought iron was the stock-in-trade of the early smith. It welded easily, worked easily, turned a satisfying black with age, and didn't rust badly. Old pieces of it show a pronounced grain like wood instead of the crystalline structure of even mild steel.

In an age of machines, mild steel became preferable to wrought iron, and except for some hand-muddled iron from California and perhaps some still found in Norway and Sweden, I know of none to be had new. Old pieces can be found in very old wagon tires, andirons, hinges — anything forged long enough ago, before the blast furnaces made uniform mild steel so cheap and available.

Master smith Steve Stokes, of Stokes of England (now based in Charlottesville, Virginia), forged this knife of Damascus steel — layers of high-carbon and mild steel welded in alternating layers, then etched and polished.

Junkyard Steel

Carbon Content	Type of Steel
Mild Steel	Most scrap (bars, rods, angle iron), chain, some nails,* some bolts.*
Medium Carbon	Most shafting, tie rods, drive shafts, reinforcing rod (not dependable), frame steel, crankshafts, piston rods, some engine valves† (many are alloys), pump sucker rod, some bolts, cable, some nails, railroad spikes, many hooks, most saws (circle, crosscut).
High Carbon	Coil and leaf springs, wrenches, torsion bars, some engine valves, hayrake teeth, picks, mattocks, railroad rail, axes.
Tool Steel	Grader blade, plow points, drill steel, machine tools, files, rasps, stone-working tools.

*Nails, spikes, horseshoes, and most bolts are low carbon, somewhere between mild and medium.

†Caution: Avoid sodium-filled engine valves — they are highly dangerous under heat or impact.

Cast iron — also called gray iron, and often found at junkyards — is not malleable. It is good for heat resistance and is used in engine blocks, stoves, and cooking pots. It should not be confused with cast steel, which is usually high carbon.

Your Own Scrap Heap

You will certainly acquire a scrap pile of odds and ends around your shop that will contain valuable old iron and steel, no matter what your wife's opinion of it. But resist the urge to scavenge just for the sake of getting it free. If it's doubtful that you'll be able to use a supply of particularly grotesque-shaped iron, leave it for the next fellow. He may think in different twists than you do.

Understand, of course, that iron may be shaped and reshaped countless times, recycled on and on. Steel may lose its temper in a fire, but it can be rehardened and tempered again, unless burned out. You will occasionally run into a piece of old, pitted iron that splits or crumbles when you try to forge it. It's called "red short," and I know of no way to use it. Other than that, most old stuff is as good as new.

We mentioned salvage yards. Garages, old barns, demolition sites of all kinds, flea markets, and your neighbor's workshop all are good sources for mild steel and carbon steel. Most machine shops and equipment maintenance shops have scrap piles out the back door, and the owner will often part with bits and pieces.

I found a steel bar that was ½ inch by 1 inch and about 4 feet long many years ago while poking around the burned remains of a late-19th-century artist's studio back in the woods. I have no idea what it was used for originally, but I forged a set of chisels and punches from it; these continue to be my basic tools, in spite of a plethora of others acquired and made since.

You will be able to find steel much easier than you can hammer it. And for most jobs you won't have to shell out for new stock after your own supply dump has grown.

4

HAMMERING IRON

I have heard several veteran blacksmiths reply to the question of what exactly they do, in this way:

"Well, you get a piece of iron hot, then you hit it with a hammer." And that's what it's all about. Iron and steel are malleable when red hot, and hammering lets the smith shape them by degrees to his purposes. We'll simplify things by referring to mild steel as iron, and carbon steel or tool steel as steel. Sometimes this gets confusing, but you need to know the difference. More than once I've shaped a piece of what I thought was mild steel, then tossed it into the slack tub to cool it, only to have it shatter in use — carbon steel in disguise. It's usually easy to tell that carbon steel is tougher when you're hammering it, but sometimes it gets past you.

I built this forge around the cast firepot shown underneath. It has a triangular clinker-breaker/tuyere, the handle of which I later bent downward to hold the proper alignment, and a trapdoor ash dump, held shut by another weight. The blower was separate, connected here with flexible exhaust pipe.

Starting the Fire

The first step is the fire in the forge. First of all, make sure the ashes are dumped and the air passage into the firepot (the tuyere) is clear. Now crumple four pages of newspaper into a tight ball and lay it or some wood shavings in the bottom of the forge, with coal and coke, if any, raked away to all sides. Light the tinder and, if you're using newspaper, turn it so the flame is at the bottom. Crank the blower or operate the bellows gently at first, as you rake in a layer of coke from the last fire, and fine coal. Increase the air blast so you don't smother the flames, then add more coke and coal. Now really get on the blower crank and pile coal, coke, or both on top, building a mound at least six inches high. This will effectively smother the fire and you'll get a dense smoke, mostly from the newspaper. Poke an airhole down through the center, which will be burning out by now, and you'll get a flame to come up, burning some of that choking smoke off.

Use the poker to push in the sides of the hollow where the newspaper burned out, and repunch the airhole. Now rake up coal from all sides to form a miniature Vesuvius and pack it down with the back of the coal shovel. Keep cranking. You'll see the smoke diminish as the coal turns to coke and the fire clears. When it's out to about a three-inch diameter, sprinkle water around the edges of it to contain it and to help steam impurities out of the coal. Use water from time to time to keep the fire where you want it. A tin can with a long handle is handy, maybe with holes punched in the bottom for a sprinkler. I use an old bullet-mold lead-melting ladle.

Be careful not to put too much water on the fire at any one time. I once cracked a cast-iron portable forge this way. Use a sprinkle only as needed when the fire starts to spread too much.

Don't put iron into the fire until the smoke clears. Raw coal should not come into contact with the iron, or you'll get melted coal tar all over it, and you won't be able to see what's going on through the smoke. To heat a long section of bar, for instance, let the fire spread out, keeping it narrow with water. And move the bar back and forth to distribute the heat.

With air blasting up from the bottom, the heated iron tends to oxidize faster near the bottom of the fire, and less at the top, where it also absorbs some carbon from the burning coke. It's best to keep the iron down in the fire instead of on top of it, to help hold the heat in. Your fire should be deep, a minimum of the six inches we mentioned, and the iron can be one or two inches down in it. As you heat iron, the core of the fire will keep burning out, so push it in from all sides and bank new coal around the outside, sprinkling a little water each time.

When you leave the fire for a few minutes, heap raw coal over it to make coke while you're gone. The fire will eventually go out, but it'll burn a lot of coal before it does. Sprinkle water on it when you're through with it to save coal, remembering to use very little water. If you want to keep it going while you do something else, a favorite practice among smiths is to shove a pine knot down into it.

More about watering: Too much water will put out the fire, and the water will combine with the sulfur that is burning out of the coal to produce sulfuric acid, which will eat away at the bottom of your forge. That in itself is a good reason not to leave the forge out in the rain, a sloppy habit at best.

Drawing Out the Iron

Now let's start with drawing out a piece of iron, a basic step in blacksmithing. Push the end of a ⅜-inch round rod about 20 inches long through the lip of the coal mound into the fire. Turn your blower steadily, but not fast for this rod, or you'll just waste coal. As you gain proficiency, you'll have two or more irons in the fire at once for higher production, but for now one is enough.

Watch the iron through the hole it made when you pushed it into the fire, or pull it out and inspect it. When it's red to about three inches from the end, pull it out and stop cranking. The tip may be white hot by now, and always remove the iron before it starts to spark. The sparks are burning iron, and much of this will pit or even burn away the rod. Any sparks at all will ruin carbon steel, since the fire eats its way down through it.

When iron starts to give off white sparks, it's burning and is too hot for anything, even forge welding, contrary to popular belief. Burned iron will pit and be weakened; burned carbon steel will crumble apart when hammered.

Dark red heat is the first color you'll notice. You can work mild steel at this temperature, but you won't get much done. Cherry red is a bright glow that will sometimes flicker as you look at it. This is the heat that's usually best for hardening steel, and any iron can be worked here. Orange or yellow is hotter, and the iron is more malleable still. Light yellow, which we call white hot, is just before the iron burns. You'll see it glisten, as the surface is actually molten at this heat.

Of course, the hotter the iron, the more you'll be able to do with it before it cools, within limits. It's a good idea to heat the iron back farther than you plan to work it so the cold iron won't draw heat from the hot part. Also, the cold anvil will cool the iron as soon as you lay it on it, so work fast. Of course, the term "striking while the iron is hot" comes from the smithy, along with "losing your temper" and "too many irons in the fire."

Scale, which is oxidation, or instant rust, will form on the iron with each heat. Hot scale will burn you but won't hurt the metal, unless you heat enough times to take several thousand layers off. Try to keep the number of heats down, especially on carbon steel, because you get pitting as well as scaling if you overdo it.

You may stand anywhere you choose in relation to the anvil, contrary to many an old-timer's belief. It isn't a good idea to impale yourself on the horn as you swing around from the forge, however. Position the anvil block so it's close to the fire but not in the way. It's generally better to crank the blower with your left hand if you're right-handed, which means letting go of the blower to grab iron or tongs as you pick up the hammer in your right hand. If you're left-handed, reverse this.

For most work, the anvil should be at knuckle height when you're standing next to it. I keep a second one four inches higher for light, close work so I don't have to bend over so much.

All right. I've left you holding that red-hot rod for too long. Get it across the anvil face with the tip quite near the far edge. To draw the iron, hit it with the hammer at a slight angle, flattening the iron and pointing it a bit. Turn the iron up on its edge while it's fairly thick to work this side, forming a square, tapering toward a point. If you flatten too much at first, it'll fold when up on edge. Keep moving the tip back from the edge of the anvil as you hammer, which will lengthen, or draw, it.

Workability by Heat Color

HEAT	FORGING ABILITY
Dark red	Malleable with some difficulty. Can work iron, but will require harder hammering. Carbon steel will surface crack at this heat.
Cherry red	More malleable. Can work iron, low-carbon steels. Quench most steels here for hardness.
Orange	Universally malleable. Can work iron, all steels, punch holes, upset.
Yellow	Easily malleable. Watch for burning. Can work iron, all steels, upset. Welding possible. When working carbon steels, cold hammer lightly while cooling to pack crystalline structure.
White or yellow-white	Welding heat — molten on surface. Watch for burning, particularly if steel or iron is deep down in fire. Cold hammer carbon steels lightly while cooling.

Turn often, which will keep the square shape still somewhat pointed. Now, if you're reducing the rod to a given size, draw the tip to that size first, then direct your hammering to the thicker part of the rod to match the end. If you're pointing, as for a nail, a hook, or a tapered loop, take the tip all the way to a point. Let the face of the hammer hang over the edge of the anvil a bit, so you won't knock dents in the anvil face. Use a lighter blow on the tip, and reheat as the red fades. Steel will often split if you hammer it thin too cold. Once pointed, watch closely in the fire; it will burn easily.

Drawing iron to a point. The flat hammer face and the flat anvil give you flat surfaces, of course, so you get a square when drawing iron. It can be rounded again by standing on the corners, hammering to an octagon, then rolling and hammering round.

A Word About Hammering

Use an elbow action, holding the handle as far toward the end as is comfortable. You'll hear admonitions to hold it only by the extreme end from other smiths, carpenters, and assorted busybodies. Hold it where it feels right and gets the job done. Period. But do rest the hammer on the anvil every few strokes. That lets your muscles relax, which is nice when you want to let go of the hammer without having to pry your fingers away. As you gain accuracy, you'll find you grip the hammer farther out toward the end.

I'm reminded of an old friend, tinsmith T. A. Hart, who was working in another man's shop some years ago, learning the trade.

"Quit choking that hammer!" the old tinner yelled at him. T. A. explained that he had better control up close the way he was doing it. The old man was unimpressed.

"In this shop you'll learn to hammer right!" he roared.

Now, T. A. has a stubborn streak, so he determined to hammer whatever way felt best, which happened to be with his hand about halfway up the handle. Finally, the old smith could stand it no longer. He grabbed the hammer from his apprentice and ran the handle through a nearby table saw, cutting off the part that wasn't getting used.

"Now at least we won't waste that extra wood," he said. T. A. took the hammer back and went to work again, quietly. This was not the reaction the old man expected from his hotheaded apprentice. T. A. hammered on.

"Didn't it bother you for me to saw your hammer in two?"

"Hell, no. That's your hammer. Cut 'em all off if you want to, you crazy old man." Still hammering.

I've put in a lot of hammer strokes, and except for the wise-mouthed codger at every crafts fair, nobody tells me how to hold a hammer. That old man is unavoidable. I sometimes hand him the hammer without a word, or offer to bet him I can hit his finger if laid on the anvil, even though I'm not holding the hammer right.

Generally, heavy blows can be directed with more accuracy from the end of the handle, light ones from farther up toward the head. And accuracy is important. I was once shaping a bowie knife blade on the anvil at a crafts fair, slamming away with a ball-peen hammer, bending close, I missed clean, and the hardened hammer came up off the hardened anvil like a shot, scoring one on me right between the eyes. The crowd loved it, probably thinking the blood running down my nose was all part of an act. Wanted me to do it again for them.

You will wear yourself out pecking lightly at heavy steel all day when you could whop it harder faster with a bigger hammer, but with a selection of them from 16 ounces or less up to 4 to 6 pounds, you can get the feel of hitting the iron effectively. The object is to shape the iron and try to retain the function of your arm in the process.

A lot of us smiths, and carpenters too, suffer from tendonitis, commonly called tennis elbow. It comes not so much from too much hammering as from too much improper hammering. It's aggravated by snatching the hammer up after each blow. Let it die on the iron, then lift it more slowly and your arm will last longer. When the iron is hot, the urge is to slam it as fast as you can, but resist this. There's always time for another heat.

"Heeling" the hammer is holding the handle too low, so the face is not parallel to the anvil face. "Toeing" it is the opposite. Proper anvil height and a proper swing, which has some push from the shoulder in it, will get the hammer down flat.

Squaring and Rounding

Back to our piece of iron. This squaring is the first step, because the hammer and anvil are flat, and anything you hammer is going to flatten. If you

Cutting Off Iron on the Hardy

The piece is cut from both sides for a more even end.

The final blow in cutting should be just alongside the hardy, to shear off the iron. This piece is a carbon-steel tooth for a rock rake.

want it round again, turn the squared rod up on its corners and hammer it eight-sided. Now roll it and tap it round. Keep hammering until it's as smooth as you want it. Old-time smiths wanted machinelike smoothness to their work, but you'll find yours is appreciated more if the hammer marks show. Why would anyone pay for handwork if it looks machine made? But don't get sloppy. There's a broad range of workmanship between slick factory precision and botchery.

Making Nails

I start students on nail making, that old apprentice beginning job that went on and on before machine-cut square nails almost 200 hundred years ago. Nail making was often a cottage industry, and folks hammered nails by the fireside in winter or at small forges set up at home.

To make nails, draw the rod to a point, leave it square, and insert it through a tapered hole in a tool called the nail bar. If the right length sticks through as it tightens up in the taper, mark a spot about ⅜ inch past the big end at the bar, on the hardy. If the nail's too short, hammer it some more. Now take the rod out of the bar, heat again, and cut it from at least two opposite sides on the hardy, almost through. Stick the rod back into the nail bar and break off the rod at the cut. Now push the rod back into the fire, put the nail point down into the pritchel hole in the anvil, and rivet a head onto the nail, quickly.

If you cut the nail from one side only on the hardy, the iron will lean over as you head it, and the head will all be to one side, like a railroad spike. Some of the ⅜ inch you left for the head will jam down into the bar, but you should have enough left. Rivet with the flat hammer, or use the ball peen for a decorative effect. Now quench, bar and all, in the slack tub, which will contract the hot nail and let it out of the bar.

That's how nails were made for thousands of years. Today hand-forged nails are chiefly decorative. I pin pothooks to beams with them, or hang my tools on them, or do a quick demonstration at a crafts fair. Once I get going I can do a nail in one heat, but there's a trick to it. The trick is placing the cut-off rod back into the fire as you head the nail, so there's still enough heat in it to draw the next nail without another heat.

Shaping Iron

Bending hot iron is done best by hanging it over the edge or the horn of the anvil and hammering the iron downward; or setting the end on the anvil, holding the other end out, and hitting it in the middle; or clamping it in the vise and bending it. Let's say you want to make that tapered point into a hook instead of a nail. Lay it across the horn near the point, hot, with about a ¼ inch over, and hammer the end down. Move more rod out at each blow, and the tip will curve down and under.

Now lay the hook on its back on the anvil face and tap the tip down to where you want it. If the bend is irregular, slip it over the point of the horn and round it up. If you want a reverse bend at the shank of the hook like a question mark, hang it, back upward, on the horn and hit it downward at the shank. If you want the tip of the hook to curve outward, hang just this tip on the horn, back downward, and hit it about a ¼ inch back from the tip, which usually requires another heat.

Shaping iron is all logic. You have to hit the iron for a while to see how each blow will affect it, but before long you'll be able to get some economy into your hammering. Heat as often as necessary. You should be able to draw a 3-inch-square point on ⅜-inch rod in one heat with some practice. I'm not as fast as some smiths, but I routinely draw and point a 2-inch hook, bend it, and mark it on the hardy for cutoff in one heat. Use a big hammer to get the job done faster.

I'll say here that most beginners are afraid of hitting the iron too hard. Remember that you can reshape any mistake you're liable to make from hitting too hard, so go ahead and slam it. There's a lot to be said for getting the basic shaping done with a big hammer while it's hot. You can finish with lighter blows later.

Decorative Loops

For a loop, as in the end of a poker, heat a longer section of the rod. For precision, mark the beginning of the loop first, with soapstone or a nick with a file or chisel, and bend a right angle on the edge of the anvil. Now turn the rod over, and start the bend as you did with the hook, over the anvil horn. With luck the tip will curl under and finally touch at the first bend.

One Method of Forming a Loop Handle

1. The stock is drawn to a point, then rounded.
2. Then the point is curved over the horn of the anvil.

3. The point is brought on around from below.

4. The piece is laid on its back to close the loop.

5. The reverse curve is hammered into the shank. An alternate method is to bend here at a right angle first, then form the loop in the bent part.

6. If a reversed tip is desired, it's put in last, then the loop is trued up.

Turner White,
a student in
one of my courses,
upsets the end
of a poker to form
an ornamental ball.
A Springfield,
Missouri, attorney,
he has a forge
at his farm.

Adjust the circle on the horn. You may want to cool just the right-angle bend a bit in the tub before you start the loop, to keep from straightening it too much, but it'll usually be near the cold iron and this won't be necessary.

Usually I don't make a loop this way. I start the loop with the point, finish it to size, then hang it on the horn and put the reverse bend in last to get the loop centered. Both these methods give you a functional, plain loop.

Now try drawing another rod to a taper, rounding it, then looping, keeping the pointed tip bent outward. Makes a nice piece of work. A variation is to hammer the tip flat but not pointed, round the corners, and make a reverse bend at the end of the loop. You will develop your own preferences. In time, who knows? Collectors will be able to identify your work by its distinctive touches.

Twisting

Twisting is a decorative operation applied to square stock, squared sections of round, or bundles of rods. Heat the entire part to be twisted by moving the rod back and forth in the fire to an even red heat. Now clamp in the vise and twist with tongs, locking pliers, or if you have a loop in the end, with a bar through. A light twist produces a spiral; more gives a tighter look.

Twisting is also a corrective measure if your fireplace poker, say, doesn't come out with its offset point parallel to the loop in the handle. Often a piece will need a twist to align it after it's finished.

Upsetting

Upsetting is a more complicated operation that thickens the iron instead of thinning it. The metal is heated, then jammed from the end to spread it. A ball on the end of a rod is a common upsetting operation. Heat the end, then hammer from this or the other end against the anvil. Watch the edges that

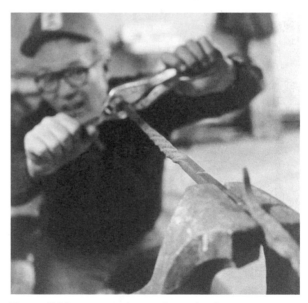

Turner White puts a decorative twist into a fireplace poker. The point is the forge-welded bird's-beak design.

will mushroom out, and keep them hammered in. Once the thin edge folds over, it must either be ground off or welded on again. And welding will thin the upset. Repeat the heating and hammering until you have a mass of metal to shape into the ball.

This is a frustrating process for beginners. Very little progress is made with each heat, since keeping the edges hammered in tends to reverse the upsetting process. Most smiths prefer to hammer a ball from larger stock, punch it, and then weld or rivet the bar into it.

I learned another method of upsetting from fellow blacksmith Bob Patrick — to bend the end of the rod at a right angle, then lay the rod on its back on the face of the anvil and upset the bent end back into the main rod. Take care here not to let the bend become a fold. This bend/upset can be repeated as often as necessary.

The basic technique of drawing hot iron can be speeded up by using the peen of the hammer first to create a series of humps, which are later hammered flat. The hammer shown here has a 45-degree peen instead of the standard straight or cross peen.

Some smiths upset the ends of rods to be welded so there'll be extra metal to compensate for the thinning by hammering the weld. I don't do this, since there are two thicknesses in the lapped joint anyway.

Right Angles

To make a simple right-angle bend, hang the iron over the edge of the anvil, hammering down and then inward to the side of the anvil, and then on top again. The iron will stretch at the outside of the bend and be rounded here. If you need a sharp right angle at the outside of the bend too, you'll need to upset as you bend.

To do this, make the basic bend, then heat and shape a convex curve out of the angle on the horn. Now place this shape back on the edge of the anvil, and hammer outward toward the bend from each side. That pushes metal toward the bend, upsetting there. Finally, shape the right angle again in the extra iron you've upset. This makes a stronger bend, since there's more metal there.

Hammering Flat

When you hammer iron, you stretch it. This is important to know for several reasons. If you plan to hammer a round or square rod into a flat piece for, say, a knife blade, your eight-inch rod may be a foot long when you're through. The narrow hammer peen can be used to direct the stretching better. Used parallel to the rod, the peen will flatten it without lengthening it so much. Used the other way, the peen will lengthen it more.

To draw any iron more quickly, use the peen across the rod, skipping spaces, to make humps. Then hammer the humps out with the face of the hammer. Once flat, you'll bend your knife blade up when you taper it to the edge. Again, the peen, used along the length, will reduce this bending. I have no doubt that the curve of knife edges is the natural shape from tapering during forging. It also makes it more efficient for slicing.

Another method for offsetting the bend when a knife blade is tapered is to bend the basic steel bar the other way before beginning the taper. Or you can stand it on edge and straighten it as you go. If you do this, watch for folding at the thin edge.

Carbon steels, such as those used for knives and chisels, must be worked hotter than mild steel, or surface cracks can result. Remember not to let the carbon steel burn, though. If it does, you must cut off that part and start over.

Tapering an iron strap for, say, a door hinge, involves drawing, which will upset and thicken the narrow part. You'll have to flatten it again to keep a uniform thickness. All this will lengthen your strap hinge a lot, so allow for it. I shape the tapered end with whatever decorative design I'm using, then cut off the wide end if it's too long when I'm through drawing it out.

This decorative end may require more iron than that left by the taper. You can upset, but that's the hard way. Leave a blob at the end as you draw to a taper, if the design requires it. You've seen the classic spade or disc at the end of strap hinges on old doors. The smith usually flattened the decoration very thin, using just the iron at the end of the strap. That's one style. I like a uniform thickness, which requires more iron.

A Three-Way Split

A split curl or a three-way split can be done with the iron left in the strap. Cut back into the end with the hot cut or the hardy (it's hard to see what you're doing on the hardy because the cut is on the underside). Finish the cut with the hot cut and the iron clamped in the vise. Cutting all the way through on the anvil will dull the hot cut or hardy, even though this may be done on the table or step of the anvil, which is softer than the face. Then bend all but one prong out of the way while you work that one to a point and curl it. Use the punch with the piece clamped in the vise to finish the cut, and bend the prongs out of the way.

A Pothook and Nail

Here's a good project to help develop control of the iron. We'll make a pothook and the nail for it all in one piece. Hammer a ⅜-inch round rod to a tapered point, then put a right-angle bend in it 2½ inches from the point, making the bend sharp at the outside. Now heat again and put the point through the nail-heading bar and head just a little. Now cut off the rod three to four inches up, heat, taper, and form the hook. These are handy hooks for your shop, where you'll be more apt to have timbers to drive them into, or for kitchens with beams sturdy enough.

The same hook can have a loop, for a forged nail to be driven through into a beam. For conventional house walls, flatten the shank and punch a hole for a screw to mount it with. Then hunt for a stud behind the drywall.

Punching holes doesn't weaken the iron like drilling them does, because most of the metal stays in, pushed out to the sides. Also, punching is faster if you have the iron hot anyway.

Lay the hot iron flat on the anvil, then quickly drive the punch into it until you're almost through; the cold anvil will cool the thin piece remaining. Now take out the punch and turn the iron over, with the hole right over the pritchel hole in the anvil. Drive the punch back through, placing it in the center of the outline you'll be able to see from the top.

The thin plug will drive out, and your hole will have a slight taper from both ends from the tapered punch. Now cool the punch in the tub, and flatten the iron again on the anvil face. If this closes the hole a bit, punch it out again. If you want a bigger hole, heat again and use a drift (that's the big, tapered punch, usually with a handle) to enlarge the hole, over either the pritchel or, if it's too big for this, the hardy hole.

Your punch will get hot in thick pieces of iron, so remove between heats and cool it. You'll need to reshape, harden, and retemper it from time to time or it will mushroom and bend. If it mushrooms down in the metal you're punching, it'll be hard to get out. We'll discuss keeping your tools in shape in the chapter on tempering.

Riveting

Riveting is a simple and effective way to join two pieces of iron. Punch holes and flatten the surfaces to be joined, then cut a piece of round rod a bit longer (⅜ inch or so) than the thicknesses of the two pieces to be joined. You should have a tight fit so the rivet will stay in place as you put the whole thing in the fire. The ends of the rivet will heat first, as you turn the piece. Now place the rivet end on the face of the anvil or on a riveting tool that has a rounded cavity in it, and use either a standard or ball-peen hammer to mushroom the top end. The bottom will spread on the anvil face, or it will round in the riveting tool, and can be shaped more with the ball of the hammer.

Making a Pothook and Nail

1. To make an integral pot- or coathook with forged nail, draw to a point first, then bend a sharp right angle in. Next, head slightly in the nail heading bar, shown here.

2. The heading bar will allow a flat surface to hammer on when mounting the hook.

3. Heat the stock, cut it off, and draw it to a point. Then bend it over the horn.

4. Complete the bend.
5. Heat and shape the tip. This makes it easier to hang a pot on, and also protects a coat from tearing on the sharp iron.
6. These hooks can be hammered into heavy beams, or into lighter wood if a lead hole is bored slightly smaller than the nail shank.

Making Your Own Tools

I always have my students make several of their own tools in the basic course I teach. Cold chisel and punch are the first, although we wait until later to harden and temper them. These are easy projects, involving only drawing and tapering, with rounding for the punch. I do stress that each piece of work should be shaped as much as possible with the hammer, leaving very little for grinding or filing.

Tongs

Now let's go on to a more complicated tool, a pair of tongs for your own use. A good basic pair for light work would be about 12 inches long with flat jaws, and close with about ½ inch still open between the handles.

Use ½-inch square stock of low- to medium-carbon steel about nine inches long. Cut two pieces and heat both in the fire at the same time. This lets you do each step on both as you go, and helps keep them uniform. Flatten 2 inches of the end of each to about ¼ inch thick, and round the corners. Now rotate the pieces 90 degrees, or to the next surface of the stock, and flatten the next 2 inches.

Now shape the ends of this second two-inch flat surface at a long angle, using a flatter if you have one, or the side of the anvil tail if you don't. A flatter is a square-faced hammerlike tool that is struck with a hammer to shape the iron with more accuracy. You'll hammer a shoulder or step here, with the outside of each piece flush or flat its entire length, and the inside stepped to half the thickness, or ¼ inch, then back. Both pieces should still be identical. Hammer the shoulders till both pieces fit together as tongs should, then mark the pinhole with soapstone or with a center punch.

Punch both holes. I use a ¼-inch pin of spring steel because a bigger hole and pin can weaken the tongs. Now try a piece of pin stock through both halves to see how they operate. Make any necessary adjustments by hammering or grinding or both, but don't rivet the pin in yet.

You still have several inches of heavy ½-inch-square handle to deal with. Heat and draw both to a long taper, which will stretch them considerably. They can be about ⅜ inch in diameter at the ends, and should be somewhat less than the full ½-inch-square, even up near the flattened part at the pivot.

A Pair of Tongs

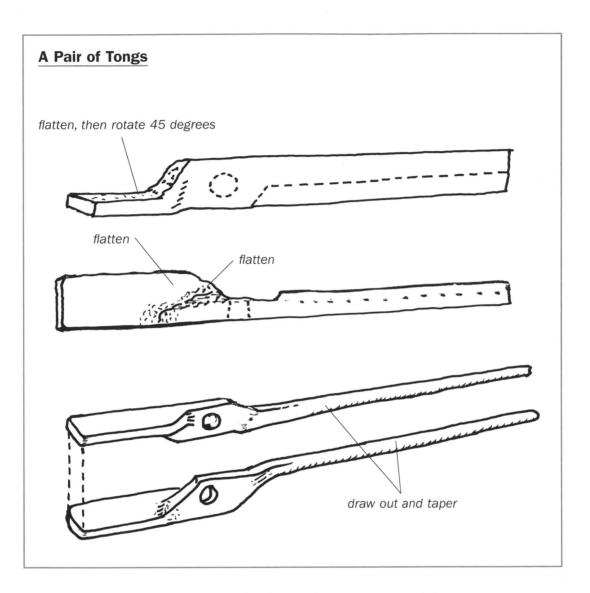

flatten, then rotate 45 degrees

flatten

flatten

draw out and taper

Round the handles to a comfortable feel for you. I give mine an oval shape, so the feel of the handle isn't too sharp.

Now cut a pin of the spring steel, insert, heat, and rivet tightly. While still hot, force the handles apart and work them till the tongs cool. Don't quench to cool with the steel pin hot in there, or it'll become brittle.

Forging a Pair of Tongs

1. John Fitch, a participant in one of my workshops, holds the ½-inch-square stock while I take a turn at the hammer. We're flattening the section where the pin will go, the jaw already having been shaped.
2. Al Lemons and I draw out the handle on a pair of tongs he's forging.

3. The small flatter allows a more exact flattening at the pin joint in the tongs.
4. This pair of tongs will have cupped jaws for holding round stock. A shouldering tool (a blunted hot chisel) is used on the jaw, set into the step of the anvil.

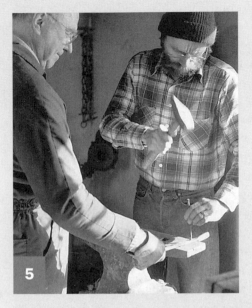

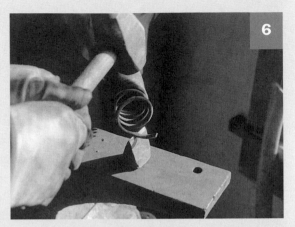

5. The flattened matching face of this half of the tongs is punched for the pin.
6. The pin is cut hot from tough spring steel stock, so that a smaller hole and pin can be used. This steel is too hard to hacksaw.

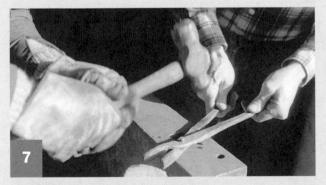

7. The pin is heated while in place, taking care not to drop it into the forge fire, then riveted tight with the ball peen hammer. The tongs are then worked while hot to free them for use.
8. Final adjusting of the jaws and handles is done in the vise after riveting the pin. Any rough edges can be ground or filed now, too.

For a blacksmith's finish on your tongs, heat to a black heat, which is just under darkest red, and dip into oil. It'll smoke and stink and maybe catch fire, so dip and remove quickly and hold downwind. The oil will penetrate the surface of the iron and turn it a nice black. Rub off excess oil with a rag as the iron cools. You may have to dip again several times if all the oil burns off the first time.

You may want to crosshatch the jaws of the tongs with a three-cornered file for better holding, but remember this will leave a pattern in the hot iron you handle with them. And you will notice a tendency for mild steel tongs to bend if you really clamp them tight, so you may want to use a better steel. I'd suggest ⅝-inch shafting, of about 30- to 50-point carbon, or maybe even reinforcing rod. These steels are tough instead of hard like higher carbon stock, and shouldn't need hardening for use as tongs.

I mentioned earlier that a favorite pair of tongs I have is made from horse hoof nippers with the jaws worn out — cheap at junkshops. Heat and straighten the jaws, and hammer till they meet with the right space open between the handles. Cut off any excess jaw on the hardy, since one will usually be longer than the other. These are high-carbon steel, so don't quench at all. They will cool in the air to a good, normalized toughness. And be sure not to hammer the jaws at less than red heat.

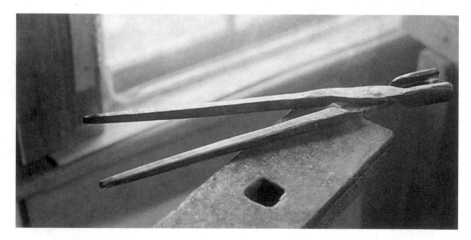

The finished tongs. A 120-grit aluminum oxide or Carborundum sanding disk can be used to smooth the rough spots, as was done on this pair. Then they were heated to black heat and oil was rubbed on them for a blacksmith's finish.

A Hot Cut

Another good learning project is forging your own hot cut, a very useful tool. We'll make it of medium-carbon steel, since it will be used to cut soft red-hot iron. A section of car axle or large reinforcing rod or shafting will do nicely, about one-inch thick.

Punch the hole first, since this is an operation that can easily ruin the piece if you get it off center, and it's better not to have a lot of work invested at this stage. I punch the hole before I cut off the stock, which means I can hold it by hand for better control. Start about three inches from the end with a small punch, cooling often, and drive almost through as for any hole. Reheat when necessary, punch through from the other side, and drift out round. When you have a hole ¾ inch in diameter, flatten the spread outside of the bar to elongate the hole.

If you have an oval drift or handle drift, drive this in and flatten against it. If not, don't try to get the outside too flat or you'll distort the hole. The hole itself should taper from each end. This is good, because a hole tapered out wide at the top only for the spreading of the handle with the wedge will leave a sharp exit hole at the bottom that will cut into the handle with long use. The double taper will let you taper the handle going into the metal, then taper it again with the wedge to hold it tight.

Now heat the bar and cut off on the hardy about three inches past the hole, cutting from opposite sides, parallel to the hole. If you have hammerhead tongs, use them to hold through the hole in the piece. If not, use round or square-grooved tongs on the end to draw the cutting edge to a taper. Leave some space between the taper and the hole so you don't distort it. Draw to an edge as you would for a cold chisel, and leave about a ¼-inch thickness near the cutting edge, tapering back gradually. I go ahead and hammer the edge now, working at the edge of the anvil or on the horn so I don't dent the anvil face. Any time you miss the work and hit the hard anvil face with the hammer, expect it to bounce up into your face.

Once the edge is shaped, true up the struck end of the tool. Now let it cool in the air or in ashes to soften the steel, or anneal it. Air cooling is called normalizing and will work all right on medium-carbon steel. A thin piece of high carbon will harden in the air even without quenching, and it will be hard to file or grind. Burying the red-hot steel in ashes or in sand will slow cooling and make the steel softer, what's called the annealed state.

Making a Hot Cut

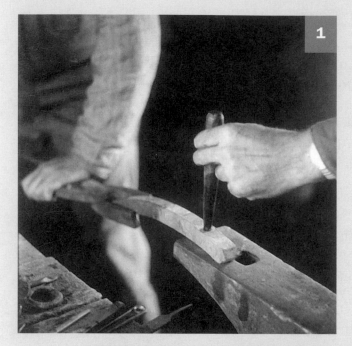

1. Punching the handle in heated stock for a hot chisel. It's easy to get the hole off center and ruin the tool, so it's done first.
2. Cutting off the hot chisel stock on the hardy. Cut in the direction of the taper to be put on the tool.

3. Tapering the tool to the cutting edge, using hammer-eye tongs.

4. Filing the final edge on the hot chisel, after annealing. Rough spots can be filed or ground out at this stage.

5. A finished hot chisel. This tool only needs to be medium-carbon steel, since it only cuts hot metal. It's heated to cherry red and quenched; its carbon content is not high enough to make it brittle.

Now grind the final shaping or, if you like shiny-smooth tools, file or polish on a belt sander. You're ready now to harden the cutting edge. If you're using axle steel or other medium carbon, just heat this edge to about an inch back, to a cherry red, and quench in water. You should be able to mark it with a file but not cut into it easily. If you've used a higher carbon steel, we'll tell how to harden and temper it in the next chapter.

The hot cut handle can be of anything, since it's used only to hold the tool in place when hit, not to swing it with. I pick up warped or otherwise imperfect seconds from handle mills or wherever I find them, two for a dollar or so for such tools. Sometimes I drawknife them from hickory or ash before the hearth on winter nights, along with the axe handles, cant hook handles, and others I use.

The heating and hardening of any tool you make will do away with the smooth shine you may have put on it during grinding or filing. If you want this on your tools, use the belt sander and perhaps a polishing wheel after hardening. The steel will rust easily, however, so keep a thin film of oil rubbed onto tools.

These are just some of the basic forging operations you'll use. We'll discuss others as we go along. You'll develop your own approach to shaping iron and steel and learn shortcuts yourself and from other smiths.

In the following chapters we'll go into more advanced techniques. In practice, stay with simple tasks until you master them, or your work will always be rough. There's nothing wrong with a primitive look to your forging, but you'll always feel like a beginner around another smith whose work shows more care and pride than yours.

5

HARDENING AND TEMPERING STEEL

No area of working with metals is so laden with misinformation, ignorance, even downright superstition, as the hardening and tempering of steel. I recall an overgrown country boy back in the 1940s who refused to file the broken tip of his pocketknife blade.

"Why not?" I asked him. We were both about 11 and I'd say equally dull, but he'd just moved into that part of central Arkansas, and I figured he knew a lot of stuff. He was, after all, bigger than I was. Sure enough, he fixed me with a look of patient disdain, and I knew I was about to learn one of life's secrets.

"Because the steel in the file will draw the steel out of the knife blade. Anybody knows that."

My estimation of this hulk dropped several notches. Had I heard right?

"The steel in the . . . Aw, c'mon, now, you don't believe that!" He did. Fervently. Some old uncle or grandpa had told him, right out of the 1800s darkness of sharecrop farm belief.

Well, I finally succeeded in persuading the kid that the file could do no such thing, using the familiar use of files on axes and saws to illustrate. A dim light began to dawn.

"Well, okay, I'll file it, I guess," and off he went, mumbling to himself. "But if it ruins the blade," he called back, "it's your fault."

More common is the belief that heat will draw the hardness from steel (which it will) and that the tool is worthless from then on (which it isn't). Steel can be rehardened and tempered indefinitely.

I know an aging blacksmith who's better at hammering iron and forge welding than I'll ever be, but he has a vast empty space in his knowledge when it comes to steel. I watched him punch holes in red-hot horseshoes, then reharden his punches (pritchels), which were of the finest high-carbon tool steel, by coating them with case-hardening compound. Every time he used the tools, he heated them, applied the compound (which raises the carbon in steel), and quenched. Maybe he was trying to make diamonds or something.

The same man told me he could temper mild steel, which is so low in carbon that quenching has no noticeable effect on it. And a book currently in print has several paragraphs on the hardening and tempering of wrought iron, which by its nature cannot be hardened.

And an old codger my brother worked with on a horse ranch once had a stallion cut, to try to improve his violent nature. Apparently the veterinarian had forgotten his scalpel and used a razor blade for the operation. The horse was still too spirited to be worth much, even as a gelding, and the old man blamed the use of the razor blade.

"Too high a metal in that razor," he swore. "Ruined that horse, usin' a razor blade on him." Lots of old-timers refer to carbon steel as being high in "metal."

Steel of a given carbon content is meant to do only limited jobs. I was at work in my village shop one day when a very old man stopped to watch.

He'd been a country smith many years ago and wanted to know if I ever sharpened plow points. Yes, I had, but there was not much need for that now.

"Well, I used to a lot," he remembered. "Draw 'em out if there was enough steel left, weld more on if there wasn't. Used to use a piece of car axle."

"On plow points?" I asked. Axle steel won't get very hard, being only medium carbon.

"Yeah, good steel, axles. And I thought I was a pretty good smith, y'see, could do just about any work."

"Well, I'm sure you were. But why'd you use axle steel on a plow point?"

"Good steel, axles. Only thing was, I never could get one really hard. Don't know yet what I done wrong."

All he did wrong, of course, was use a steel that couldn't do the job. Car axles must take impact and twisting, and are made from tough, flexible steel, not hard steel. The abrasion of soil and rocks on a plow point requires a high-carbon steel, like grader blade or jackhammer stem, or a piece of another plow point.

The Basics of Tempering

Two of the most important parts of blacksmithing are choosing the right metal for the right tool or piece of equipment, and (if you're working with carbon steel) tempering that metal properly for the job the tool will be used for. "Tempering" just means regulating the hardness in a metal by controlled heating. Metals of different carbon contents need to be heated to different temperatures to achieve a particular level of hardness, and the amount of carbon in the steel determines the potential hardness. As the steel is heated in the forge, it will turn a series of colors to indicate the temperature the metal has reached.

Hardening the Steel

The tempering process begins with hardening the metal, or heating it to its optimum hardening temperature (usually indicated by cherry red), then cooling it quickly. The carbon in steel makes it harden when cooled suddenly, so quenching the red-hot metal in water will make it as hard and brittle as it can be made. If allowed to cool in the air from the red-hot stage, the steel will be "normalized," or relatively soft. If cooled slowly in hot ashes or in a dying forge fire, it will be "annealed," or as soft and workable cold as it can be made. This is true for most steel, except for sophisticated alloys. Also, some specialty steels, like hard-surfacing welding rod, are designed to harden without quenching.

As I said, the quenched hardness depends on the amount of carbon in the steel. A car axle made from medium-carbon steel, for example, will never become as hard as a jackhammer stem or a plow point, which are made from high-carbon steel.

Tempering

Now, a glass-hard tool (except for maybe a file) is just about worthless, so the hardness has to be lessened to the right degree, or "tempered." You don't temper steel to harden it; you harden it first, then temper to soften to the right toughness for the particular job to be done. Tempering steel hard is like slamming a door open, as blacksmith Wayne Haymes says.

Heat applied to the quenched, hardened steel softens it, making it tough (or slightly flexible) instead of brittle. The degree of hardness or toughness is determined by both the original carbon content and the degree of tempering heat applied. Sometimes the entire piece of steel is tempered; sometimes just part of it is. Usually, the working edge is hardest, and the rest (particularly an end to be struck with a hammer) is softened through tempering more.

Tempering Colors

After the steel is hardened then quenched completely, the surface of the metal has a dull, blackened covering of coal dust and iron rustlike skim. Shine the dark surface of the metal with a grinder (so you can see the colors) and lay it back in the fire. Heat slowly from the bottom until tempering colors begin to appear on top. This is so the heat can soak completely through the piece before you see it. A straw color means the least amount of brittleness has been removed. You'll see this begin to appear at around 450°F. Bronze follows, with more heat, to take away more hardness/brittleness. Next is purple, then blue, often called peacock blue, the softest of the basic tempering colors.

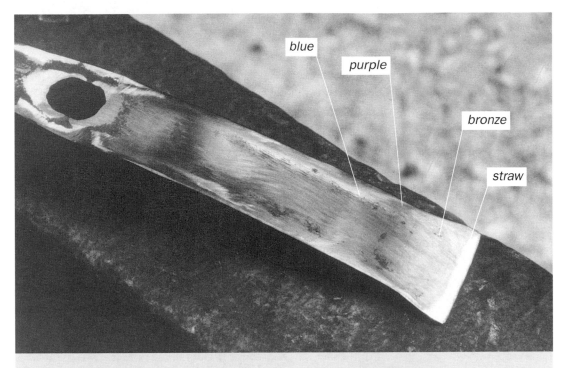

blue

purple

bronze

straw

The tempering colors shown on this forged stonemason's hammer indicate the relative hardness of the steel. In this case, the flat striking face and the chisel end both need to be the hardest, or straw color, while the handle hole area must remain the gunmetal color, normalized for toughness. The tempering process must be done in two separate steps, concentrating on one end of the hammer at a time. After the whole hammer head is brought to red heat and quenched, the striking face is tempered. Then the top of the hammer is ground shiny to enable the blacksmith to see the tempering colors run clearly toward the chisel point. Heat for the tempering process is applied carefully toward the chisel end to avoid nullifying the previous process on the striking face. The straw color means less heat has reached the chisel point so that it remains the hardest.

A slightly softer bronze follows. Next is purple, then blue. Past the blue is a gunmetal color, indicating that the most heat has been applied here and resoftened the steel, leaving it tough rather than hard. (See the back cover of this book for a color photo of the running tempering colors from the chisel end.)

Tempering can be done in the forge, but you may find it difficult to control tempering heat. If you like, use a propane or acetylene torch to draw temper after you've hardened the steel in the forge and quenched it. You can even temper over a gas kitchen-stove flame. Instead of the 1,500 to 3,000°F of the forge, you need only 450 to just over 600°F for tempering, once the steel is quenched. Remember to heat from the opposite face you're watching for colors. This means the heat has come all the way through, and gives a truer picture of the state of the steel inside.

Hardening and Tempering, Step by Step

Now let's detail the hardening of a piece of 100-point-carbon steel, which is one percent carbon by content (in other words, a high-carbon piece of steel). The metal could come from any of several items — automotive spring steel, old hammer or axe heads, or hayrake teeth — but for now it's just a piece of steel.

Reaching the Optimum Quench Temperature

Start by heating the steel to the optimum quench temperature. Usually, this is cherry red, or 1,400°F. Watch for a pulsing glow when the metal is held in the shade. That means the molecules are tearing around like mad in there, and the steel is at optimum quench temperature — the point of decalescence. You can see this color better if your smithy is semidark. Don't worry too much about the pulsing; some people never see it.

A more exact way to tell when the point of decalescence has been reached is by using a graded quench. On the hardy, cut partway into a hot bar of the steel, at intervals about an inch apart. Then heat from near white heat at one end, spreading the colors to dull red over several notches. Quench the bar in water. Now break off the first section, which will be quite brittle, by hanging it over the edge of the anvil and hitting it with the hammer. Remember the colors you heated to at the various notches. (Maybe you'd better write them down.)

You'll see a coarse crystalline structure to the steel at the white-hot break, getting finer as you approach the breaks at cherry red or near it. The color that gives you the finest, most satiny crystals at the break indicates the point of decalescence. That's the color you should use when

Tempering Steel

Tempering is the process of heat-treating or softening hardened steel to a certain level for a certain use.

1. Choose a piece of steel to be worked for a certain purpose by its carbon content.

2. Forge the steel into the shape you need.

3. Heat the whole piece to red heat.

4. Quench. The whole piece is now glass brittle.

5. Grind a shine onto the surface of the portion of the piece to be tempered.

6. Return to the forge and carefully reheat the area to be tempered, watching the shined surface for the appearance of the four tempering colors, which indicate temperature. The straw color, the first color seen, is coolest and leaves the quenched steel nearest its brittle state. Bronze indicates a slightly softer state. Purple is next. Blue is the hottest color, leaving the steel nearest to normalized, or soft. After blue, the steel goes back to the gunmetal color, indicating that all hardness has been removed. At this point, if you have not tempered the portion of the piece the way you need to, you have lost your temper and must quench it (reharden) and reheat for the proper colors.

7. When tempering colors start running, remove the steel partially or completely from the forge and let the proper color run to a stop at the proper place. DO NOT QUENCH.

quenching this type of steel. It won't always be cherry red. And cherry red to you won't be cherry red to me or to somebody else. But we'll use that designation to simplify things.

Keep in mind that steel tends to weaken when heated above its quench temperature (the coarse crystals don't bond to each other as well). So whenever you work it hot, as in welding, most smiths agree you weaken it. You can compact the structure of the steel and toughen it again by hammering it (lightly) as it cools from the high heat, down past quench temperature. Don't hit it hard much below cherry red, or you can cause surface cracks. Light cold-hammering helps toughen the structure.

Quenching

Dunk the quench-temperature steel into the tub, to cool it as fast as possible. If it's a thick chunk, clear water is all right to quench in, from my experience. Other smiths won't quench thick steel in water, fearing contraction of the surface while the center is still fully expanded, causing cracking. I've never had this happen. If the piece is thin, it's liable to crack in plain water. Again, some smiths use brine to slow the cooling and avoid cracking (although other smiths believe that it *speeds* the cooling. Pick your belief; I don't use brine at all). You can slow the cooling somewhat by using hot water, however.

An oil quench will definitely slow cooling, avoid cracking, and result in a tougher steel that's not quite as hard as what comes out of a water quench. Old motor oil is fine, although it will probably catch fire. (And smoke. I said you'd get dirty. All oil is flammable, unfortunately, but you can cover it with a lid to smother it.) You don't get maximum hardness quenching in oil, but sometimes it's a matter of keeping your piece *in* one piece. Do move the steel around in the liquid for faster cooling.

Now the steel is as hard as it can be — and 100-point steel can get hard. You won't be able to mark it with a file, and trying won't do the file any good. This hardness is too much for anything except maybe the cutting edge for a stone tool, or another file. And the hardness also makes the piece brittle, so now we'll temper it.

Running the Colors

Shine the steel with a grinder or emery cloth, and lay the steel back in the forge, over a slow fire. Eventually you'll notice a yellow glaze spreading.

The color, an oxidation, indicates temperature only (it indicates hardness only in relation to the carbon in the steel and how quickly it was cooled in the quench).

This yellow, or straw, will be followed by bronze as more heat is applied, followed by purple and then blue, with some shades in between. After blue there'll be a noncolor we call black heat, followed again by red if you leave it in the fire long enough. This whole process is called tempering, or "running the colors."

The straw glaze means we've softened the steel just a little. This is still good for stone tools; the steel is less likely to break. It is good for metalworking tools, too, such as drill bits or metal lathe cutting bits. Next is bronze — softer, but in 100-point steel, still hard. This is good for cold chisels and drills, where you want maximum hardness but still some safeguard against shattering. Next is purple, better for wood chisels, punches, and some knives (if they don't have to bend), and for impact tools that don't need as hard an edge but get jolted a lot. (Tools that really get slammed around resist breakage better if they're of lower-carbon steel, by nature tough instead of hard.)

The last of the oxidized colors is blue, softer yet, and still less apt to break. This is good for axes, knives, saws, hay hooks, fishing gigs, and springs. I temper a broadaxe to blue if it's 100-point steel, because it slices along the wood grain. Felling axes I leave at purple, because the edge tends to turn (from chopping against the grain) when tempered to blue. Also a wire edge (the thin sliver of steel that sometimes results from sharpening) is more apt to form if the axe is tempered all the way to blue.

Past the blues, you've taken most of the hardness out and are close to normalized steel, which, at 100 points, is good for lots of things, too. A kitchen knife of this steel is plenty hard when normalized, and so are some springs and tools for working soft wood. There isn't much point in using such a high-carbon steel here, though. It still tends to be more brittle than tough, and a lower carbon content steel tempered to a color farther up the scale for about the same hardness would be better.

Remove the steel from the fire. You stop the softening process at the proper color by taking the steel out of the fire a little before it reaches the color you want it, since it will have a lot of heat left in it. If you've moved it around in the forge for even heat, you have an evenly tempered piece of steel. It'll tend to heat faster at thin places and at the edges. If colors run too fast, you can slow them by a quick dip in the tub, and your color will stop and won't change until heat reaches that color at that spot again.

But be careful of this. Too many blacksmithing guides tell you to temper to color, then quench the whole piece to stop it. Such directions are an example of not enough information to serve well in actual practice. But if you're heating a punch, say, from the struck end, and get the bronze you may want at the tip, the other end will be much hotter, even red, and quenching to stop the bronze will make the struck end brittle, obviously.

Do apply tempering heat slowly. The heat must soak into the metal completely to be effective, and a quickly run temper will only affect the surface or thin parts of the tool, leaving the inside brittle.

Tips and Troubleshooting

If the colors change too fast, and you wanted a bronze but the steel goes to blue before you can stop it, heat to red and quench again, shine, and repeat the tempering. When the steel is the right color, you can quench quickly and then pull it out to stop it without hardening a hotter part of it too much, if you're very careful. We'll talk more about this later.

Remember that, contrary to just about every set of directions in print, the tempering color chart doesn't tell you a thing about the state of the tempering unless you know the carbon content of the steel, and unless you've heated it to optimum quench temperature in the proper medium (cold water, hot or salt water, or oil).

And a piece of lower-carbon steel, if used in the experiment we just outlined, would change all the results in relation to colors. For example, 50-point quenched steel might have to be drawn not to a blue for an axe, but to a straw in order to be hard enough to hold an edge. The 50-point steel would be tougher, too, not having that extra carbon in it for more hardening potential.

If you don't know the carbon content of the steel you're working with, quench it from red hot and purposely draw the temper ("temper drawing" is the complete name of the process) to a color showing softer temper, to ensure against breakage. You can reharden and draw to a harder color as use dictates, but if it's too hard at first, you'll break the tool and have to start over again.

Tempering an Axe

Now let's apply this procedure to a practical tool — say, an axe made of, or resteeled with, 90-point carbon (0.9 percent). (Often old axes and other tools you'll find have been in a fire, for one reason or another — in a tool-shed or house that burned down — and the temper will be gone.) Most automotive leaf spring steel, that old standby, is close to this. Heat the axe head to cherry red in the forge. It's not necessary to heat the entire axe, just a couple of inches from the cutting edge back, which will be the working part. We're assuming the axe was normalized or annealed from whatever forgework was done on it before. Get it hot slowly and watch for that flickering glow as it heats. If you don't see it, quench at what looks like cherry red to you. Quenched in clear water, the axe will be hardest; in oil, softer but tougher. Let's quench this one in oil, which will guard against cracks as the quickly cooling metal contracts.

Now, don't even drop the axe while brittle or it might break. Use emery cloth or a grinder to shine a place from the edge back up the side of the tool. If you had a heat-treating oven and pyrometer (what professional heat-treating operations use to get the proper hardness), you wouldn't have to do this, but you have the forge and your eye. Lay the shiny axe in the forge over a low fire, or hold it up over the fire to heat it slowly, from the handle hole outward toward the edge.

You'll see the light yellow or straw color spread over the shiny area toward the cutting edge. This will be followed by bronze, purple, and then blue. Heat slowly so these colors move out gradually. Rapid heating will cause a compact color band, hard on one side and soft on the other; with this combination, the axe will soon be sharpened through to soft steel. You want a band about two inches wide from straw to blue. That way, it'll take years of sharpening to get through the hardened steel, when you can re-temper or resteel the axe if necessary.

The colors are caused by oxidation on the surface of the metal, and only show up on the part you've ground down to clean steel. You can see them best in shade or low light. Stop the tempering as the first blue starts to replace the purple at the edge.

Ideally, as we said, you should take the tool out of the fire ahead of time so the color runs out slowly, to stop at the right place by itself. You'll be able to do this easily in time, but you can lay the axe on the cold anvil to slow the run, or dip it quickly in and out of the slack tub; this is a bit tricky.

I harden and temper tools such as axes in one operation. With practice, you will learn how to heat more than the area it's necessary to quench. Now quench just the two inches or so of the working-edge area of the axe, leaving considerable heat back up toward the handle eye. Move the tool up and down in the tub to avoid a hot/cold line, which would almost certainly cause a break. Grind shiny with a stone, before the remaining heat can run down into the part you've quenched. Use the heat left in the steel to run the colors, just as you would in quenching totally and reheating. This won't work with a tool to be tempered to an even hardness, but most tools should be softer and tougher back from their hard cutting edges, to avoid breakage.

Now, that same axe, tempered to blue, would be too soft if made of 30-point shaft steel, too hard if made of 120-point file tool steel. You'd quench most shaft steel and draw maybe to a straw for the right hardness, or maybe not at all. You'd probably have to normalize the file steel in the air to keep it from breaking in use. Axes and saws take a lot of impact, distortion, and twisting in use, and must be softer than other cutting tools.

Proving Your Temper

The proof of your temper is in the use of the tool. You may well find that the axe we've been discussing chips in use, which means it's too hard at the edge. Or the edge may turn in use, meaning it's too soft. A lot of variables are at work here, beginning with just how much carbon is really in the steel. Assuming you've been able to determine that fact, look next at the quench temperature. Cherry red is pretty bright, but I've seen smiths quench at a dark red instead, which doesn't produce as much hardness. I call cherry red that which is brightest, but with no discernible orange in it. Again, practice heating steel and watching for that pulsing phenomenon. If and when you detect it, use that color for that steel when quenching.

If you've done everything as well as you can and the axe is still too soft, reheat and quench in cold water. Put the axe cutting-edge down into the water quickly, and move it around. I've never had an axe blade crack in water, although I've been warned that it can happen. This will give a harder quench and, subsequently, a harder temper at blue.

Lastly, if the axe is too hard or not hard enough after you've done all this, go to a higher or lower color: purple if the axe is too soft, past blue to gunmetal if it's too hard. Chances are it won't be too hard at blue, but you may have a maverick piece of steel. With the current recycling of everything from the junkyard into new steel, you could have a couple of ball bearings in there.

The same 90-point spring steel, made into a wood chisel, would be drawn to a bronze or a purple, harder than the axe because it's less apt to break, being placed against the wood and struck on its softer end, or on a wooden handle with a mallet. The axe, swung often against hard knots and seasoned wood, sustains some twists and jolts that require the steel to be tempered so soft, and it's a rare axe that will hold a razor edge for long.

I have such an axe, which I steeled with spring steel and drew to purple for a harder edge. It has never cracked but is too hard to file easily, so I use a sharpening stone on it. It doesn't need sharpening often, but I confess I fear that someday I'll chip it.

One of the mistakes modern smiths make is the same one the old-timers made — they get hung up on the tempering colors and treat the tempering process as if all steel were of the same carbon content. I've had other metalworkers ask me what color to temper a hardy, say. Of course, there's no set answer. A hardy of axle steel is not a hardy of jackhammer-drill steel, so the colors would be quite different for the same hardness. I happen to have a hardy made of each of these steels. The one made from the axle I just quenched; it wasn't temper drawn at all. The other is drawn to a purple. Accounts of early smiths tell definitely what color an axe should be, or a knife or a drill bit, with no consideration of the steel itself. And this is simply because they didn't know that the carbon content had that direct a bearing on the results.

The plain truth is, you can run beautiful tempering colors on a piece of unhardened steel, and they won't mean a thing. You can also run the same impressive colors on a piece of mild steel or even wrought iron, which can't be either hardened or tempered. The colors, remember, only indicate temperature, and that is meaningful only if everything beforehand has been taken into consideration.

You wonder how the old smiths, and even the modern ones, get close with their tempering. Well, in time those who work a lot with carbon steel soon notice the differences and allow for them. Unfortunately, few of these craftsmen have written about the finer points of tempering for the guidance of the rest of us.

Bending the Rules

In all fairness, there is a lot of slack here. A tool made of a steel within 10 or 20 points carbon of another won't be a great deal different in use. Certainly, by fudging tempering colors, you can get almost the same final results.

Take the two examples we've just discussed, the experimental chunk of 100-point steel and the 90-point axe. These were close enough in carbon content to need approximately the same treatment for a not-too-different tool. An axe of 100-point steel would be a bit harder to file than one made of the 90-point steel if both were drawn to blue, and the 100-point axe would hold an edge longer. But on a bitter January day, chopping a white-oak knot, a chunk might break off the harder steel (this happened to me with a too-hard broadaxe a few winters ago).

Also, the steels available to the early smiths were less sophisticated than those today, and the demands of their customers less exacting. After all, there weren't factory-made tools of the same monotonous near-hardness available for comparison.

That old steel might have had hard spots and soft places in it, and a host of other irregularities. Heating and hammering tended to spread and smooth out the rocky spots, giving more uniformity to it. So a belief still exists today that hammering iron toughens it. In the strictest sense, this is not so. While lengthening the fibers, or "grain," of wrought iron by hammering toughens it, if the carbon was not in there, no amount of banging on it would put it in. Now, peening or hammering carbon steel will alter the crystalline structure of it, packing it and making it dense. This also relieves stresses in forged steel, and I often cold hammer a knife blade or other tool for this purpose. And, as we discussed earlier, hammering high-heated steel as it cools will toughen it.

Each smith learns to bend this or that rule a bit too, for the desired result, but he may not be able to tell exactly how he did it. Seldom will you heat, quench, and temper a piece of steel evenly throughout, or it would all be pretty simple. This is usually what happens to hardware-store tools.

Ideally, you'll want your tools hard at the working edge, tough through the shank, and softer at the struck end, if any, to prevent chipping or breaking. This requires the more complicated hardening and tempering processes.

One-Heat Tempering

I often harden and temper tools with just one heat, as we discussed with the axe example. Here's another example — a cold chisel. Quench to about an inch, then dip deeper in and out, keeping the inch of the cutting end in the tub. While there's still some red heat up the shank, remove, lay across the anvil, and rub quickly with a hone or whetstone to shine (emery used here will burn you). Heat will bleed down to run the tempering colors toward the cutting end. If these stop too soon, heat more; if they move too fast, slow with another quick dip in the tub or by laying it on a cold part of the anvil. This is not a good practice on very thin punches, because you get a normalized state for most of the length of the shank, and a thin punch will bend if used on thick metal.

By the two-stage process of hardening and then tempering, you will be able to control tempering colors better; a wider band means more toughness farther up the shank. Remember, you don't want hardness at the struck end, or steel will fly.

One-heat tempering takes practice, since the colors can run all the way off while you're shining the tool. And, of course, you can cool so much there won't be enough heat left to temper. The thing to watch most closely, though, is that tendency for a cold/hot line at the quench line. Spread it out.

I seldom try one-step tempering on such tools as double-bit axes or, of course, with tools on which I want an even temper throughout. Until you get the hang of it (and remember that some very good smiths never do), stay with complete quenching and a separate heat to temper.

Tempering a Knife

A difficult bit of tempering is a knife blade, partly because it's so thin, partly because, to get an even temper, concentrating the heat in the right places is quite exacting. We'll talk more about knives in the chapter on tools for your homestead, but let's temper a knife here.

Tempering a Knife Blade

1. A carbon-steel paring-knife blade is filed to final shape while annealed, or softened by cooling slowly from cherry red.

2. After heating to cherry red again, the blade is quenched in oil to harden it, the first step in the heat-treating process.

3. The brittle, hardened blade, of 95-point carbon steel, is shined with emery cloth or a hone so tempering colors can be seen. Then heat from a very low forge fire draws the hardness, or tempers it to the proper degree, shown in oxidized colors. For the actual tempering, the blade is held in tongs and moved continuously in the heat for even tempering.

Knife makers usually work cold, grinding their shapes from flat stock, then heat treating evenly in their ovens for a uniform, almost-hard quality. They seek new and sophisticated alloy steels to allow for toughness, hardness, and corrosion resistance, all with a minimum of heat-treating effort. Most of them are loud in their condemnation of plain carbon steel for knives. However, my late friend Frank Waite, a master knifesmith, created beautiful, serviceable knives from new spring steel stock, as well as from the exotic alloys.

Carbon steel, hardened and tempered to the use for which the knife is intended, remains, after thousands of years, still the finest cutting edge we know. (I'm sure more than one metallurgist will disagree.)

Let's use 50-point carbon steel, hot hammered to shape. You will have annealed the blade in hot ashes, to cool it over a period of an hour or more, to its softest. Now you do whatever grinding and filing is necessary to remove humps, pits, and forging marks (if you want them removed). If you've forged carefully, there will be a minimum of cold work necessary. I draw file, pulling the file toward me for better control. Drill handle-mounting holes now, too, or thread the tang, which is the tapered end the handle goes on, if you plan to use a hollow handle and shaped nut to hold it on.

A vegetable chopping knife my wife uses. It's of normalized spring steel, with forge marks left in.

Heating and Quenching

The blade will be so pretty now you won't want to put it back into the fire, but do so, moving it for uniform heat. Watch the tip and cutting edges; they heat faster. You can grip the blade by the tip with the tongs to slow heating there. Heat to cherry red, then quench, edge straight down, in oil. Period. Again, some smiths use brine, with enough salt to float a potato or an egg, but others insist this won't work. And clear water is risky. My nephew once spent four hours hammering a skinning knife from rare steel, then quenched in clear water. The knife cracked right across. Contracted too fast. Anyway, the oil will slow the cooling. You can quench in water by dipping in and out several times to slow the cooling, but you have to know just how. Hot water, too, will slow the cooling. Hell, just do it in oil.

All right, your shiny blade is gray now, and hard. And there shouldn't be any warp to it if you dunked it straight down. There might be some warp if you hammered too much on one side and not enough on the other in forging. If so, we'll straighten it later. Fifty-point steel is medium hardness, so you could conceivably leave the blade here. The chances are good that you'd break it, though. Shine it again with emery cloth, or on a belt sander if you have one, all over, making sure you don't get it hot enough to start colors running (around 450°F).

Tempering

Now, 50-point steel is a tough metal, not a really hard one. You will temper to different stages for different uses. If after some use you find the knife too soft or too hard, you can remove the handle and reharden and retemper it. For general use I would draw to a bronze first. If this proves too soft, harden again and draw to light yellow or straw.

As we said earlier, it's relatively easy to temper evenly throughout, as most knifesmiths do in heat-treating ovens. In the forge you keep the blade edge up, moving over a low fire, watching to keep more heat on the thicker spine of the blade. The thin edge and point will go through the entire color range while you're waiting for straw on the back, or the spine of the knife if you're not careful. Remove from the fire often, to slow the heating and for closer inspection. You may have to reharden and temper several times to get it right. Don't hurry, and banish onlookers during this process.

Tempering Colors*/Heat Treatment

Color	5 pt. (mild)	40 pt. (medium carbon)
Annealed	No effect	Soft, like mild. Drill, grind, sharpen tools before heat-treating. Use like mild.
Normalized	No effect	Soft but tough. Drill, grind, sharpen before heat treating. Use like mild; stronger.
Blue	No effect	Tough; will bend. Still relatively soft.
Purple	No effect	Tough. Tongs, triangle gongs. Tougher than mild.
Bronze	No effect	Tough. Prybars, axes, saws, knives, shafting, splitting mauls, froes.
Straw	No effect	Semi-hard. Knives, wood carving tools, drawknives, hot chisels.
Quenched	No effect	Hard. Hardies, cold chisels, punches, wood carving tools, augers.

and Their Effects on Steels for Typical Uses

100 pt. (high carbon)	150 pt. (tool steel)
Soft but tough. Drill, grind, sharpen before heat treating.	Semi-soft, tough. Drill, grind, sharpen before heat treating.
Semi-tough. Kitchen knives, some springs, dinner gongs, tongs.	Tough. Hot chisels, drifts, forming pins.
Tough, semi-hard. Axes, splitting mauls, knives, froes, most springs.	Hard. Hardies, cold chisels, punches. More brittle than lower carbon.
Hard. Knives, some wood carving tools, hot chisels, drifts.	Hard. Stone tools, drill bits, machine tools, plow points.
Hard. Hardies, cold chisels, punches, wood carving tools, augers, etc. A fairly safe temper for most tools.	Hard. Drill bits, machine tools, stone tools, files, rasps, plow points, some wear surfaces.
Hard. Stone tools, tools that should hold an edge but sustain no impact or bending.	Brittle. Files, some wear surfaces.
Brittle. Only use edge on stone if rest of tool is tempered.	Too brittle for use.

Remember, colors mean nothing unless carbon content is considered.

More than one smith I know has burned up a good knife blade while answering a question. It is a testament to the self-control of blacksmiths that more gabby onlookers are not bruised severely for disrupting the tempering process.

Heat very slowly, to spread the color band wide. A bronze at the cutting edge would be perhaps purple-blue at the spine. Let these colors run to a stop on their own. Dipping in water or oil to slow the colors will change the state of things enough to get you in trouble here, so I advise against it until you know what you're doing. Specifically, dipping will set the color run back, and new heat will run behind the band you stopped, but will not be visible as colors until it passes the previous band. Setting back the band by dipping, then honing a new shiny area so you can see the new colors, will not reharden a spot already softened or tempered. Got that?

The thin blade will cool quickly in the air, so your colors should stop soon after you remove it from the fire. I once hammered spring steel into a large kitchen knife for chopping vegetables. Experience had told me to normalize this steel for this type of knife, so I let it air cool from red hammering heat. Later I decided to take off a bit more steel to allow for a longer handle. I clamped the blade into the vise to hacksaw it, and succeeded in dulling a new two-dollar blade. Moral — carbon steel is tough and relatively hard even without heat treating.

Once your knife is tempered, be careful about overheating it. It's true that it can be rehardened and tempered again and again, but this will mean removing the handle each time, and more close work. When grinding the blade, wet often to cool, and watch for those same tempering colors, which can come from the heat generated by the stone. Theoretically, if your tool is tempered to blue, you can let yellow, bronze, and purple appear while grinding, without changing the hardness. But grinding should be done more carefully anyway, and there's no point in letting it heat the tool that much.

Words of Advice

Your punches and hot cuts and probably drifts and chisels will get dull, mushroomed at the struck end, and generally out of shape from frequent contact with red-hot iron. You can grind off flaring ends, or heat and hammer them back. Then reharden and retemper. Hammers themselves may mushroom with a lot of use, and they present a special tempering problem. The edges tend to harden, leaving a soft spot in the center of the hammer face. Harden under a stream of water, as from a hose, directed right into the face of the hammer. My nephew once hardened the face of a hammer nicely using a long icicle. Then draw a little temper at the edges only. It's hard to get the face too brittle, since it's the edges that tend to chip. If you make a hammer, use high-carbon steel, or at the softest, medium carbon of shafting or axle.

Test stray or untried steel before you forge it, by at least heating to red, quenching, and trying a file on it, if not the graded quench we discussed earlier. I once hammered a link to join the hook to the body of a chain binder (boomer) and made the weld nicely, as a demonstration to a class. I cooled the work in the slack tub so everybody could inspect the weld, and everything was fine. But the first time I used the boomer, the link snapped clean, opposite (not at) my weld. I'd picked up a piece of carbon steel because it had been the right size. If I hadn't quenched it, the normalized steel could have been tough and strong. But I did and it became brittle, albeit nicely welded. The break came away from the welded point from the high heat (coarse crystalline structure). My hammering at the weld as the steel cooled toughened it at that point.

There's a bit more to this lesson. I was cinching a chain around my antique crawler tractor with the boomer when the link broke. I was getting ready to haul it on my log truck over some really bad mountain roads. I could have dumped the Cat off on somebody quite easily if that piece of steel had held only an hour longer.

The same thing happened with scrap bars I forged into a pot rack for my wife. I'd put a decorative curl at the ends of two bars, then dipped them to cool for another handhold in forging. One wasn't quite nice enough, so I tapped it on the anvil horn cold for just a bit more bend. The whole curl broke off.

There are several points to remember about hardening steels in general. Do draw the temper immediately after hardening, since with high-carbon steel, any good jolt can crack or break it in its quenched state. And hardening and tempering need to be extended only a short distance back from the working edge for most tools, leaving the rest normalized. And let tempering heat soak slowly through the metal. A quick run of color will temper the surface but leave the core of the tool hard enough to break. A student of mine made a punch that was a thing of beauty, polishing it on the belt sander to a shine, tempering to just the right colors. But he ran his colors on the surface he was heating, and the punch snapped an inch from the end in use.

Tempering steel is one of the most exacting jobs a blacksmith must do; forge welding is another.

6

FORGE WELDING

Those photographs of colorful blacksmith types sending showers of sparks from steel hammered on the anvil are showing one of the most demanding tricks of the trade. Forge welding is about as difficult as tempering steel, although I know a lot more smiths who can weld than I know who can temper properly.

Joining two pieces of iron or steel by welding is simply heating to a temperature at which the surfaces of the pieces are molten, then placing them together and hammering them into one piece. And sometimes it works that simply. I recall the story told me at a Mason's meeting I spoke at. One of the older men there remembered trying to get an old blacksmith to teach the trade to Civilian Conservation Corps workers during the Depression.

"Don't know what I could teach 'em," the old blacksmith said, spitting tobacco juice reflectively.

"Well, teach them to make things, fix things. Teach them to weld."

"Don't know how I'd teach 'em to make stuff," he shook his head dubiously. "Nothin' to it, just make it."

"How about welding? That's certainly difficult. You could teach that."

"On'y thing I know to do is to get the iron hotter'n hell and beat the devil out of it," he said. "Can't figger how to teach that."

And he was about right. Of all the aspects of smithing, forge welding seems the hardest for students to get, and do consistently. When hardening and tempering are explained logically, students usually pick it up fairly well, but welding is another thing. And it's really no wonder. Taken step by step, the procedure has a lot of ways to go wrong. A lot of things must be right before that heat and that hammer can make two pieces one.

Factors to Consider

The thing that rattles most smiths about welding is the time factor. Faced with a task that must be performed in seconds if at all, a lot of folks panic.

There's the steel. Wrought iron worked easily and welded easily in the old days. Even steel welded to it easily. The mild or carbon steels of today seem to oxidize faster and are harder to weld. Also, steel burns easier at the high temperatures required for welding, fed by the blast of oxygen-laden air.

Then there's the coal. Dirty coal — that with high sulfur, that which leaves you with large clinkers — won't allow a decent weld. The crude metal clinkers melted out give off oxides, I'm told, that inhibit welding. And that is doubtless the reason, since an old blacksmith's practical joke was to visit another smith and drop a copper penny down into his forge when his back was turned. He wouldn't be able to weld all day, and of course wouldn't know why. Bits of stray nonferrous metal in the forge will keep you from getting a weld today, too.

The pieces to be welded must be shaped so that, when hammered together, the oxidation scale between them is driven out, leaving clean iron to flow together. This shaping is called scarfing, and the simplest weld is a lap scarf, in which one piece is tapered to overlap the other, which is also tapered. The resulting bond is thicker than either piece, and shaping the welded joint reduces it to size. Some smiths first upset the ends for still more iron to work with after the weld is made. This is principally because there is a widespread belief that one must pound the metal unmercifully in order to get a weld. In fact, if a relatively moderate series of hammer blows won't do it, there's something wrong, and no amount of beating on the metal, thus thinning it, will do it.

The buildup of oxidation scale, which can be reduced by the use of a welding flux, probably prevents more forge welds than any other single factor. Some dexterous smiths have developed a lightning-fast swipe of the white-hot iron surfaces with a wire brush just before the surfaces are welded. Others tap the iron against the anvil to knock scale off. If the scarf is properly made (and the coal is clean and the fire deep and the heat right, and if some other factors are favorable), the scale is effectively driven out with the first hammer blow.

Making a Chain Link

A chain link is a simple beginning weld to make, since you're welding one piece of iron back to itself. It'll hold itself in line, properly heated, instead of requiring some sleight-of-hand to get it all together on the anvil one-handed while you grope for your hammer, all before it cools in the second or two you have to work.

Scarf the ends of the rod, which should be ⅜-inch round stock, for starters. A seven-inch-long piece will make a nice, long link but still give you room to work. Bend into an oval, with the scarfed ends overlapping maybe ⅜ inch. The scarf is more than a simple taper; it should be rounded so that the iron will touch in the center first, to drive out the scale better. Now use welding tongs, something with a cutout in the jaws to hold the link opposite the joint, and heat to red. Sprinkle flux on to inhibit oxidation. Borax is the base of most welding fluxes; some smiths use borax and sand. Some use just sand. Some sprinkle iron filings in, too. The melted flux flows over the surface of the iron, sealing it from oxygen with a crude kind of glass as it heats.

Now back into the fire, which should be a deep, clean one — no clinkers. You should clean the fire before welding, even if it means taking it completely apart and building another one. With a good grade of coal and a clinker breaker in the firepot, a twist of the handle will clear the fire. Then repack it and mound coke up high. Bring the joint to white heat fairly quickly, since slow heat will heat the entire link, and it'll spread when you hit it, instead of welding. And heat fairly high in the fire to reduce oxidation. Watch for a very light yellow, which is as white as the iron will get without burning. You can cheat a little with mild steel and leave it in the fire till the first white sparks come off. That means the iron is beginning to

Blacksmith Wayne Haymes shows a student chain welding. The weld has just been made, and the joint is being shaped on the horn.

This forge welding picture shows why a leather apron and even a hat are important. Those streaks are molten iron drops and bits of white-hot scale.

burn, and a little of this won't hurt much. Carbon steel must never be allowed to burn even a little or the piece is ruined; it crumbles apart.

All right. You're probably right-handed, so you're turning the blower with your left hand. Shift the tongs to your left hand, take the link out, turn to the anvil where your hammer should be, lay the link flat down on the face as you're swinging the hammer, and hit it quickly three or four times — not too hard, but firmly. Sparks will shower as droplets of molten iron spatter and hot scale flies (wearing a hat and a leather apron is a good idea). You'll see the joint bond by the second stroke or so, and then you can hang the link on the horn to shape the weld while it's still hot. If the weld didn't happen, don't waste more hits. You can reflux and heat again if you didn't distort the scarf too much. If you did, reshape both ends and try again.

Your chain link will still be egg-shaped after welding, so shape it with the weld at the end. That way, part of each end goes around and it holds better. Commercial chain is usually welded electrically at the sides, and that's where it breaks. We'll discuss chain more in the chapter on homesteading tools.

A Few Side Notes

I weld at a very high heat, just a few degrees short of burning, and have fair success. I know smiths who can and do weld at what I'd call an orange heat, and do it well. The secret (one of them) is to have everything in place so you don't waste time and heat.

The late Shad Heller of Branson, Missouri, taught me to forge weld, and his advice was to walk through it cold a few times, to know exactly where the hammer was, just how far to zip the iron to the anvil, and so on. It helped a lot. One freak happening in courses I've taught is that of a rattled student picking up the hammer backward, and slamming the weld with the peen. Now I keep a double-face hammer to teach welding with.

Welding chain is a good beginning project. Here, a student inspects a forge weld in a chain he made.

A hoop weld, showing finishing and shaping of the thickened weld.

One of my students went on to become the smith at a historic attraction, and became a good ironworker. He had trouble with his welding, though, and finally called me to come see what he was doing wrong.

Well, his coal was the best I've ever used. His forge and tools were mostly new, and good. I watched him shape a chain link, and he did it just as I'd taught him. His heat was right, too, but he laid the link on the anvil, tapped it once lightly to sort of get it located, then took his time getting on it with the hammer.

Well, of course the cold anvil face takes the heat out very quickly. If possible, you should even hold the iron up off it until you're actually swinging the hammer. We corrected that hammering and he made one good weld after another. Tourists gathered around, so I had him show off by welding separate pieces. Then I had him weld a bundle of rods (this always looks impressive). He did that, too. As far as I know, he's doing well, years later.

Other Welds to Try

A hoop is another good beginning weld, made on the anvil horn. Scarf the ends, and leave a little tension in to hold the joint together. Some smiths even punch holes and rivet first, to help get a good weld. Old-timers used to pride themselves on their smooth welds, challenging anyone to find the joint in a welded wagon tire. They often used more than one heat, and of course were working with wrought iron.

The Eye of a Froe
A tricky weld to make, and one always harder than it looks, is the eye of a froe, that shake-riving tool of antiquity. I've seen lots of old specimens doubled for the length of the blade, and I know it's not really evidence of top craftsmanship; that smith had trouble with the weld, too. You see, it's actually easier to shape the eye, double the thickness of the blade, and start the weld at the tip, progressing back toward the eye. Simply welding near the eye is harder, especially with high-carbon steel, which burns easily.

This is a tool that should be struck only with a wooden mallet, and it need not be high carbon, but most smiths make their froes from leaf spring, which is apt to burn near welding heat. Flattened axle steel welds easier, and makes a tough enough froe if you want to put in the work required.

A Froe-Eye Weld

1. Bending the eye of a froe in leaf spring steel on the anvil horn. This tool is used to split or rive shakes or boards from straight-grained wood.
2. The tricky froe-eye weld can be made easier by riveting beforehand, using the ball-peen hammer. Or the froe can be used nicely with just the riveted joint.

3. The froe-eye weld, made after fluxing with borax at red heat, then heating to white welding heat carefully, turning often so as not to burn the steel.

4. A finished froe. This one is not hand forged; it's one my grandfather bought many years ago. It's drop forged, a factory process in which the heated steel is stamped to shape under power in a die.

Anyway, the problem is that the steel around the loop and the thin scarf edge tend to get hotter and burn before the heavy body of the blade gets to welding heat. Keep the body down, turning only after it's up to heat, and watch that the edge of the eye doesn't burn. Hammer the end of the scarf first, before the thin steel can cool, then work back toward the eye with very rapid blows, aimed first at the centerline of the weld. That helps drive out the scale. There's nothing wrong with taking a second heat for this weld, which is usually about a two-by-three-inch area.

I've seen other ways around this weld. One of the most effective, but just as hard, I'd say, is splitting the steel to form both halves of the eye and

Blacksmithing student Bill Jennings of Branson, Missouri, makes the weld at the eye of a froe.

Steeling a double-bit axe. The weld has been made, spreading the blade.
Here we're shaping it back. Later we welded steel onto the other edge.

hoop welding at the back on the horn. Some smiths punch a hole in the blade body, shape a tenon (a rounded protrusion, like an attached rivet) on the end of the eye loop, and rivet it through. Some don't even weld the eye at all. Others punch matching holes and rivet the lap before welding. This technique will help you get this weld, but remember to leave the joint open out from the riveted center, to get flux in and scale out.

You probably won't weld many froes, since there's so little good timber for split shakes these days. But I do a good trade in froes, both at fairs and as special orders. I suspect that most of them never rive a shake but are included in old tool collections, where they're bragged about as being definitely hand forged and lied about as being old.

Recently, I saw a froe that had been made in an old country shop years ago, and that had a nice touch to it. The smith evidently didn't like the fact that, no matter how sharp the bend from the eye to the blade, a little triangle is always left open. He'd laid a wedge of iron in there and welded it along with the lap weld. Nice.

If you weld any high-carbon steel such as spring steel for a froe or new steel on an axe, keep hammering as it cools down past quench temperature. This will pack the steel's crystalline structure, which will have become coarse and weakened from the high welding heat.

The Lap Weld

Graduating from welding a single piece back on itself to welding two separate pieces is a big step. Unless you rivet first, a lap weld means that the two pieces are loose until you weld them, and they're hard to manage without a helper. You heat them side by side, keeping the heat even in both, then bring them out and lay the one in your hammer hand face up on the anvil, tipped up so the cold face won't contact the iron. Then push it down with the other piece, face down, as you swing the hammer. If your heat is right, you can touch the two together in the air and they'll stick slightly, letting you have a fraction of time to position them before they go on the anvil. If the pieces are long enough to hold by hand, it's easier. Any forging operation is harder if you have to hold the iron with tongs.

A trip-hammer is handy here. You can hold the pieces together and let the machine pound them together, controlling it with your foot.

Tips and Tricks

As I said earlier, it's easier to weld if you've managed to get the joint held together somehow beforehand, as with the rivet. Splitting one end to enclose the other is another way. I steel an axe head by shaping the carbon steel bit to a V, which encloses the old axe edge. Then I use a center punch on the hot steel to make little indentations to help hold the steel in place for the weld. I flux before I drive the new steel over the old.

The early smiths, having little steel, did it the opposite way when they made an axe. They folded a piece of wrought iron around a handle shape, then welded it double up to the edge, leaving a V for the steel bit, which went inside. With spring leaf so plentiful, I find it easier the other way. Just lap welding steel onto a worn axe is a hard weld to make. I suppose riveting might help here, too, but I have never done it.

The part of the weld joint that touches the anvil will lose heat quickly and usually leave a crease where the thin edge of the scarf got too cold to flow. You can minimize this by getting the first, telling blows on one side, then flipping the piece over quickly to smooth the other side. It still won't flow together, but it will be less noticeable. You can take a second heat on this side.

A light piece of iron loses its heat quickly, so you have proportionately less time to work. Heavier pieces hold heat longer. Old forged anvils were welded twice, with two strikers, and sometimes more, alternating sledgehammer blows. One weld joined the two sections of the anvil body at the waist; the other joined the hard steel plate that formed the face.

Old books on blacksmithing caution welders to align the grain in iron before welding. That was more important in wrought iron than it is in modern steel, since the fiber of the iron definitely ran lengthwise. The grain of mild and carbon steel is less obvious, these being more crystalline in nature. But it's still a good idea. Some old iron is especially prone to splitting when forged, which can be the result of rust getting into the grain.

It's safe to say that no welded joint is as strong as the iron on either side; this was an axiom in the old days. With arc welding, which uses an electrode (a steel rod coated with flux), or with acetylene welding, which utilizes a flame and a steel rod, you can build up more metal than is on either side for strength, but a forge weld depends on those two faces hammered together, and that's all.

Weaker still than the scarf weld are the butt weld, in which the two ends of pieces are hammered together, and the jump weld, with an end hammered onto the middle of another piece. Wherever possible, you should forge from a single piece of iron and not depend on a weld or a joint of any kind, particularly at a stress point.

All those neat drawings of complicated welds in old blacksmithing guides are nice, and there was no practical way to join those pieces except by forge welding. But trying some of them with today's materials will be an exercise in frustration.

The Biggest Forge Weld Ever

Shad Heller told me the story of what is supposedly the biggest forge weld on record. It seems that a steamboat broke a six-inch-square paddle-wheel drive shaft on the Mississippi River near Memphis many years ago, and the captain wanted the drive shaft welded. No ordinary smithy could handle it, of course, but the owner of a large shop finally agreed to try. The two halves were slung from chain hoists overhead, so they could pivot from two forges to the giant anvil. They were scarfed, heated, fluxed, and then brought to white heat, as four strikers, positioned exactly, waited. Then the balanced shaft pieces were swung from the forges to lap just right on the anvil, and the strikers rained blows in perfect time, making the weld. The shaft held on the trip upriver, where it was supposedly replaced.

Alternatives to Welding

For those who never get the knack of welding, there are lots of ways around it. A chain hook can be of bar stock, or my old standby, a railroad spike, with the hole punched in instead of looped and welded. In ornamental work, a rivet often joins quite well, without the awkwardness of getting the piece into the fire or the distortion of hammering the weld. A band, too, applied hot to cold pieces, will contract on cooling to hold well. And often a small punched hole can be drifted out and enlarged on the anvil horn to a loop as large as the amount of metal will allow, letting you avoid a weld that way.

I will seldom weld an axle or drive shaft, reasoning that the stress that broke it will do so again. With steel as abundant as it is now, it's safer to replace shafting. Similarly, when I'm faced with welding steel onto a plow point or digging tool, I'd rather forge a new one. Unless the shape is complicated, the new piece is as quickly shaped and is stronger.

I will say this about forge welding: Compared to arc or acetylene, it is often a superior weld. First of all, it is impossible to forge weld without preheating both pieces, which makes any weld better. It's common to arc weld carbon steel and have it break right next to the weld. Preheating, welding with the right rod, and heat treating will stop this. Acetylene welding allows better preheating, but again the rod used to join the pieces must be of the proper carbon content for a strong weld. With the forge weld, there's no rod, so you work with the iron you have.

Ideally, any joint to be welded in any way should be heated at least to red beforehand, welded, then cooled very slowly. If made of carbon steel, the piece should be hammered while cooling, then hardened and tempered again to its use.

The Fascination of Forge Welding

Welding will continue to be fascinating to most folks. I recall a man telling me of an old smith back in the mountains who was a real wizard at his craft. Seems he'd been off to the Smithsonian Institution or somewhere to demonstrate, according to my informant.

"You know, that old man can even forge weld," he told me, in evident wonderment. (This man didn't know that I, too, was a blacksmith. He'd have thought I was too young anyway, merely middle-aged.)

Well, he finally mentioned the old codger's name, and I recognized it as belonging to a colorful smith of the horseshoe variety who looked a lot better than he was. In fact, this same old man had asked another smith, a friend of mine, to teach him forge welding just a year or so before, probably so he could demonstrate it for those historians.

But I didn't let on. "A dying art," I agreed. "Too bad he couldn't teach some of that knowledge to somebody who'd carry it on."

"Nobody's interested these days," my man snorted. "Anyways, that forge welding's too hard. Why, it took that old man most of his life to learn that. I just imagine that secret will die with him."

"Probably will."

7

TOOLS FOR BUILDING

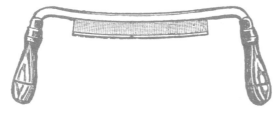

Being a builder and restorer of hewn-log houses, I consider a crosscut saw, axe, froe, broadaxe, auger, and drawknife the kinds of building tools that are essential. Of these, I have forged all but the crosscut saw. And for my kind of construction, the saw is perhaps the least essential, anyway.

Whether you plan to build with hewn logs, stone, post and beam, or conventional stud wall, your forge can help fill your toolshed with handmade or rebuilt tools that you'll treasure and get good use from. Let's look at a proposed building project — a timber-frame barn — and the tools you'll need to build it.

Three useful building tools. 1. broadaxe. **2.** drawknife. **3.** froe. All of these were in constant demand by early builders. The axe was the most essential pioneeer building tool. The drawknife is a smoothing, finishing tool for both building timbers and furniture woods. The froe split wood blocks into roofing shakes.

Rebuilding a Pick

First of all, there'll be the site preparation, if only some leveling of the dirt floor. An old, worn pick head can be had at a junkshop, often for a dollar or two. Draw out the pick head and sharpen it, then quench and temper draw to a dark straw, so the tips won't break. Then find yourself a handle. I drawknife mine from shagbark hickory or ash. Making a pick or shovel from scratch is the sort of thing that'll keep you from ever getting to work on your barn, so save that work for tools you really can't find cheap. And when looking over old pick heads, find one with as much metal left as you can. You can forge weld new steel onto it — using, say, large coil-spring steel — but it's a tough weld to make, in a place that takes a lot of stress.

The Stone Hammer

Once level, your site may need stone shaped a bit for the foundation. I make an odd-looking stone-dressing hammer by wrapping a piece of high-carbon steel (like a heavy truck leaf spring) around a handle drift, forge welding it, drawing to an edge, then serrating while it is annealed. Use a large three-cornered file to make the teeth. This may be too much for you to tackle, so shop around for a stone hammer, or use a stone chisel. (The stone hammer is really just a chisel with a handle.) If you forge a hammer or a chisel, barely draw the temper from the cutting edge; stop the straw color as it reaches there. Do draw temper from the rest, to a normalized state at the struck end (if the tool is a chisel). A really hard edge backed closely by tougher steel will handle stone best.

A good small stone hammer can be shaped from an old hammerhead. Without the hole to punch for the handle, it's a lot easier. A large ball peen will do, and you can either draw out the ball to a short chisel shape, as in a cross-peen hammer, or draw the striking face out to a longer one, shaping the ball to a flat striking face. Put a slight curve into the chisel shape, since your swing is in an arc. Leave an edge somewhat like that of a cold chisel, since the tool is used to chip or break stone, rather than cut it, and should be a little blunt. Draw the temper to a straw from the eye toward both the working face and the edge, letting it go to the face, with the entire color band stretched out about 1½ inches. This is a bit of a trick if you do both

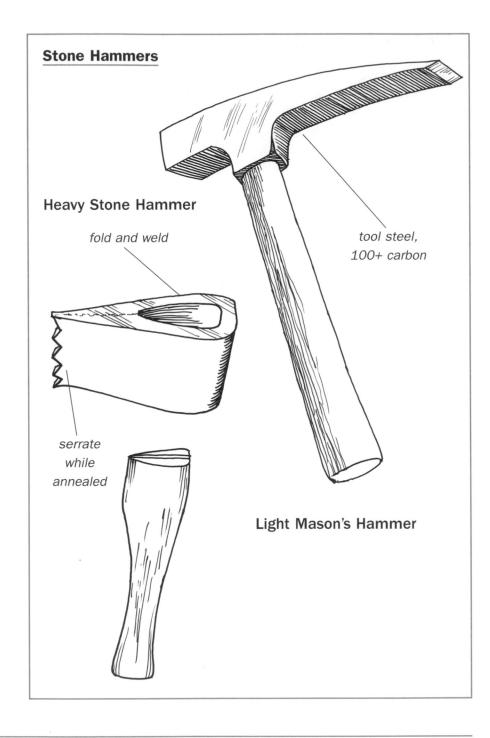

Stone Hammers

Heavy Stone Hammer

fold and weld

*tool steel,
100+ carbon*

*serrate
while
annealed*

Light Mason's Hammer

at the same time. You can dip one end to keep it cool while running the colors on the other, then reheat and reverse. The stone hammer should show colors from gunmetal at the handle hole to blue within 1½ inches from both working ends, then purple, bronze, and the straw at the edges.

The Broadaxe

Now back to the barn. If you plan to hew beams, you may not be able to find a broadaxe you can afford. They are collectors' items just now, and I rarely see one that costs less than $75. An adze is cheaper, but this is a finishing tool and won't do much toward squaring round logs.

I have forged new steel onto the cutting edges of worn broadaxes, and made them completely from plate steel. Let's say you've found an old axe with most of the metal sharpened away. You can weld a piece of leaf spring onto the cutting edge, draw out, sharpen, quench, and temper it, and have a usable axe — probably for the rest of your life.

As we discussed in the last chapter, steeling axes is not an easy weld to make. I suggest folding the new steel to leave a slot for the old cutting edge to fit into, then welding. If forge welding isn't one of your smithing skills, or if you can't find clean coal, then torch or arc weld the new steel on. Heat both pieces of steel first. Or you can butt weld, after grinding the old edge blunt, leaving a slight V for the arc welder. If you forge weld, you won't be able to get all that steel hot at once, so weld a couple of inches at a heat. Wire-brush the scale off between heats, and reflux each time.

Tempering a Broadaxe

One thing to watch for in tempering a broadaxe is warping. This seems to be a result of hammering more on one side than the other, or of uneven heating. If you use a steel of 80 to 100 points, you may find that simply air cooling or normalizing it will give you all the hardness you need. The broadaxe is used to slice off wood with the grain, as opposed to chopping across it, as with a timber axe, so it can be softer and still do the job.

If your broadaxe, of 95-point steel drawn to a blue, does warp, you can cold hammer it straight, but do it carefully. Lay the axe head flat on the anvil face, if the axe head will lie flat (there's usually a slight arch across the anvil face from side to side), or on a curved surface such as grader blade or railroad rail if it won't. Hammer evenly all over the convex surface — lightly at first, then heavier if the warp isn't coming out. Then switch to a flat surface and continue. Don't do this with steel hardened to a brittle stage. You can also cold hammer life back into sagging spring leaves this way.

Forging new steel onto worn edges also works well with single or double-bit timber axes. I have one of each I've reclaimed this way, and have done several for others. A better steel for high-impact work is automotive axle, but the hammering takes a lot of time, unless you have a trip-hammer. You can, of course, go to a steelyard and buy some thin bars of 30- to 50-point carbon steel, just right for the job.

A completed McRaven broadaxe, with some forge marks still showing. The head is mild steel, wrapped around and welded, with a poll added. The carbon-steel edge, tempered to blue, is of 95-point steel. The axe is handled for right-handed use.

Steeling an Axe

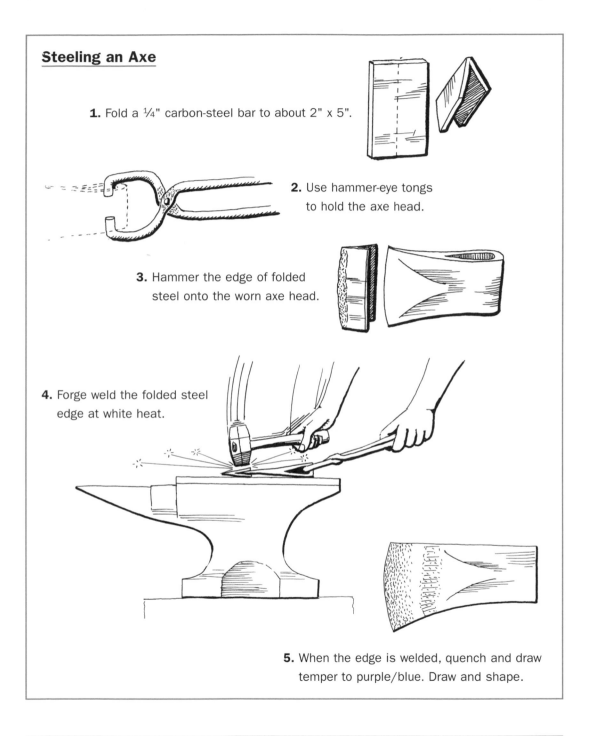

1. Fold a ¼" carbon-steel bar to about 2" x 5".

2. Use hammer-eye tongs to hold the axe head.

3. Hammer the edge of folded steel onto the worn axe head.

4. Forge weld the folded steel edge at white heat.

5. When the edge is welded, quench and draw temper to purple/blue. Draw and shape.

I'm reminded of that blacksmithing student of mine who wanted to make a bowie knife that he could throw without fear of breaking it. Partly to give him a conversation piece, I had him cut five inches of axle steel and handed him the hammer. Many hours later, he'd reduced that 1½-inch bar to a beautiful 11-inch blade, draw filed it to perfection, and will probably pass it on to his grandsons, along with the story.

Chisels

With your barn beams hewn, you'll need to mortise and pin them into place. Woodworking chisels are easy to forge and certainly more satisfying to use than the plastic-handled commercial variety. I use coil-spring steel, temper drawn to a purple. Leaf spring is good, if you have a shear to cut it to size. I also make chisels from shaft stock, tempering to bronze, and from large circle-saw blades, ¼ inch thick or more.

For sets of chisels, use spring or shaft bars of a suitable diameter for each. A ⅝-inch bar will give you plenty of steel for a 1-inch mortising chisel; I use ¼-inch bar for sculptor's finishing chisels. As with any carbon steel, hammer it hot but don't burn it. I use a shouldering tool or the edge of the anvil to step down to the handle tang, then fit a heavy, square-hole washer to it for the wooden handle to seat against. Quench and temper draw from the tang to a purple at the edge if using spring steel, bronze if shaft or circle-saw steel, stretching your colors out to at least a one-inch-wide band. I notch the tang to hold better, drill the handle with a slightly undersize bit, and drive it on dry. You may want to use epoxy glue, too. The handle itself is, for my tools, perhaps a rare piece of tiger-stripe hickory, or maple, or elm. Sometimes I will use a hardwood inclined to split, like Osage orange, and band it with iron or brass. The mallet should be hardwood; dogwood or persimmon are favorites in my region.

Drawknives

I often adze or drawknife beams to smooth them after hewing. A drawknife is a tool with so many uses that you should have several sizes. A 12-inch cutting edge is handy, although I've made them as small as 4 inches and up to 18 inches.

A split leaf spring is the simplest steel to use; cut on the shear to about a one-inch width. If none is handy, I sometimes use coil spring, which takes a lot of hammering. Leave two inches or so of steel at each end for the tangs, and hammer the tool straight. Hammering the cutting edge will curve your drawknife, and leave some of that curve in; it will slice better than a straight edge. Now draw the tangs, allowing about eight inches for the bends and handles.

After I've quenched the tool, I draw evenly to a bronze or purple. Then I draw all the hardness from the tangs, because they tend to break at the bends if hard.

I leave the tangs square and drill the handles through, to fit a square-hole washer, which I rivet over the tang ends. The square shape helps keep the handles from twisting when driven tight onto the tangs.

I use round-pole rafters in some of my buildings, and the drawknife is handy to peel the bark. It's also useful in making tool handles, tapering split shakes, and just about any smoothing job.

Augers

To drill holes for the wooden pins, or trunnels, you'll need a set of augers or bits for a hand brace. These can be made in the forge, but they are really cheap at junkshops. To forge them, use a medium- to high-carbon steel — again, spring steel is good — and hammer it uniformly flat, to about an ⅛-inch thickness. You can find

This drawknife was made of medium-carbon shaft steel, heated and quenched only. The relatively low carbon content kept it from becoming brittle and produced a tool of toughness with just this one-step process.

T-Bar Auger

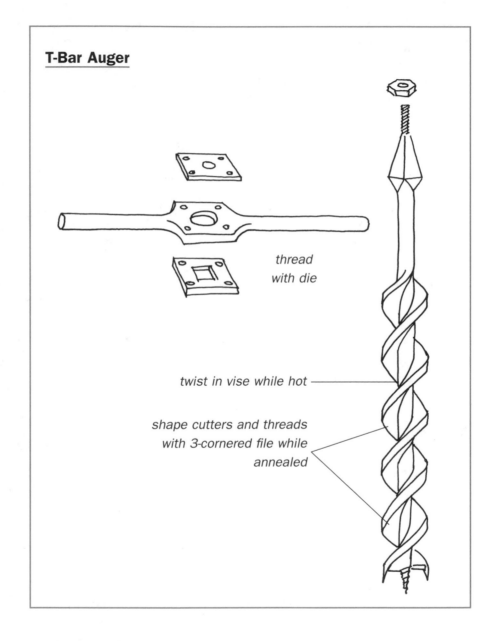

thread
with die

twist in vise while hot ———

shape cutters and threads
with 3-cornered file while
annealed

spring steel leaves this thin and cut them with the shear to the width desired; a one-inch width results in a one-inch auger. Leave a shank, hammered to a square taper.

Twisting the auger is one of those delightfully simple processes, involving hot iron and a vise, that always seem to come out right. The trick is to heat uniformly. Be sure to straighten any bends the twisting might produce. File your cutting edges while the steel is annealed, and thread the tapered point with a three-cornered file. This is a close operation, so take your time. Heat to cherry, quench in oil, and draw to purple. The bit should be soft enough to file easily, so temper to a softer state first. If you've used medium-carbon steel, such as circle-saw or shaft steel, temper to bronze.

I like the T-bar auger for large holes, because you can really get leverage on it. The brace is fast for small work, but if you're boring through, say, a white-oak top plate and the hewn log beneath it, making a 20-inch hole 2 inches across, you need torque. I shape a metal plate to take the square-tapered shank of the bit, recessing this plate into the handle and fastening with screws to it. At the top another plate with a smaller square hole is recessed, screwed down, and the shank end riveted over it. Bits get wobbly and fall out if they're not anchored tightly. Hickory and ash are my favorite woods for auger handles.

Nails and Froes

With your rafters pegged into place, you're ready to slat and shingle your barn. While I often forge nails for decorative purposes, I don't advise investing the time to do it for general construction. I do often forge extra-long heavy ones to pin the rafters to the plates. Sometimes these are a foot long, of medium-carbon steel such as reinforcing rod, sometimes started in smaller drilled holes to keep from splitting the rafters.

Nail down the slats or laths the shakes go on, with good old 10d common nails. But you will want to rive your shakes if you can find good wood that is straight, with no knots. That's often a disappointing search these days, but now and then some good second-growth red oak, white oak, or even large cedar will be available.

So you need the froe. This is a very easy tool to make if you don't care how it looks; it's harder if you do. Use heavy spring leaf, hammer it to an edge, and loop an eye in one end for a handle. Let it cool normally in the air. That, with a handle, will give you a useful tool for splitting shakes or slats. If you want to make it authentic, you'll have to make that weld we talked about earlier.

There's yet another way around the froe-eye weld if you want to take the time. I'm a believer in one-piece forging wherever possible — shaping the iron without joints of any kind. You can punch a hole near the end of the froe, then drift it out larger, and finally shape it around a piece of shafting, axle, or the anvil horn, twisting the blade one quarter-turn for the finished froe. It takes a lot of hammering, and the sides of the eye will be thin and not as wide. It's the sort of project you might want to try once, to see whether it works for you.

Finishing Touches

Barns need hinges and latches for doors, hooks and heavy nails to hang things from, and chain to lift things with. Chain making is the blacksmith's stock-in-trade at a crafts fair. It's fast, requires enough skill to look good, and is a bona fide blacksmithing activity. And it's good practice. But it's really not economical as a daily chore unless you count your time for nothing.

Hooks are another matter, since they're expensive and somehow usually absent from the scrap of chain you found along the road or bought at the flea market. A hook should be much thicker than the chain it's attached to, since the hook is open and can be pulled straight with less force than it takes to break the chain. I sometimes punch the hole that links the hook to the chain. A loop and forge weld here is not difficult, either. The thing to remember is to shape the hook before you bend it, be it a curved slip hook or a slotted grab hook.

Of course, other building jobs would require other tools, not all of which you'd want to make at the forge. I suppose I could hammer out a circle-saw blade if need be, but times will have to become much harder before I try it. The same is true of hammers. I have forged hammers of several types, but the truth is that you can pick up whatever you need quite cheaply at the junkshop.

Bars for Wrecking and Digging

A wrecking bar is a handy tool that can also be had cheaply, but it is quickly made at the forge. I use coil spring steel from ½ inch to 1 inch in diameter, heated and straightened. Flatten both ends to chisel points, then split one end with the hot chisel, working from one side to leave a sharp tapered edge. Bend the curved end toward the open taper of the split. Quench and draw to a purple or blue. A harder temper will break if you hammer on it — and you will, eventually. Put a slight reverse bend a couple of inches from the other end. Two or three of these tools, from one foot to three feet long, will be useful in any kind of building or dismantling. My brother has a giant one that's 4 feet long of 1⅛-inch steel for really big jobs.

A simple bar, but just as useful, is a straight one with a chisel tip that is flat on one side and angled on the other, for prying. I use two of these a lot in stonework. One is 3 feet long, of ¾-inch round spring steel, and the other is 1¼-inch square stock, 5 feet long. The big one is medium carbon, about like axle steel, and is quenched only. It's handy for digging or prying stones out of the ground and for lifting whole buildings or mired trucks.

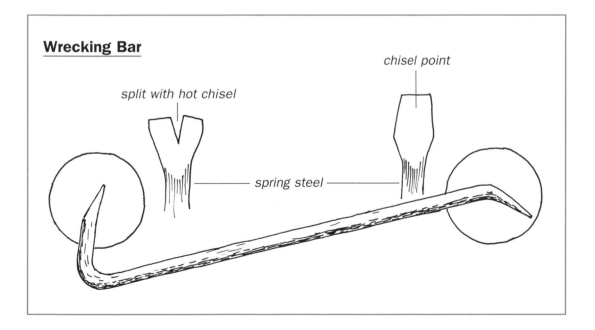

Wrecking Bar

split with hot chisel

chisel point

spring steel

If you're attempting a hand-dug well or septic tank, this big digging bar is good to use instead of a pick, for which there is often no room to swing. The bar has enough mass to shatter pretty big rocks that are in your way.

Since automotive axles are too short for a good digging bar, look in junkyards for old shafting, which usually has about the same carbon content.

Hammering the chisel point on a heavy digging bar of 1⅛-inch steel. After the initial shaping is done, we switch to the anvil horn to avoid denting the face with the edge of the hammer. This medium-carbon steel is quenched from cherry-red heat to produce a tough (but not brittle) edge without drawing the temper.

My brother John's wrecking bar, forged of 1⅛-inch steel, 4 feet long, for obstinate jobs.

This is tough steel, as opposed to hard steel, and is good for high-impact uses. A favorite in the old days was Model T Ford drive shafting. Old mills often had line shafting overhead, from which pulleys and belts drove machinery, and this can still be found in varying sizes. I have a pile of square reinforcing bar from a 1920s overpass recently demolished, and it seems uniformly usable. Modern rebar hasn't proven reliable, but you can often get usable tools from it.

Cant Hooks

For any sort of log work, you'll need a cant hook. These are usually cheap at junkshops but are so simple to make we'll discuss it here. The band is mild steel, about ¼ inch by 2 inches, punched and drifted for about a ½-inch pin.

The hook should be at least medium carbon, and the length varies. I use a small hook, 12 inches or so, for most cabin logs and logs for beams and joists. Then I have a larger one of 18 inches or so for the big stuff. Often two of us will handle a log, with the big hook on the butt end and the small one at the other.

The iron parts of a cant hook for handling logs. The hook is shaft steel flattened to a ½-inch thickness, tempered to a straw color for toughness. The handle strap is old wagon tire.

Hook stock can be coil spring steel, sheared heavy leaf spring of at least a ½-inch thickness, or small shafting. I usually start with coil of ¾- to 1-inch thickness and hammer it flat to form a hook ½ to ¾ inch thick. Upset some metal at the hook angle for more strength, and hammer a 3-inch point. Quench and temper to purple or blue.

Pin with a tough steel pin or hardened bolt, since wear will make the hook floppy, and you want this tool to bite every time. A cant hook's slipping can put several hundred pounds of log onto your body, sometimes from up on raised skid poles, and that hurts.

Another band for the end of the handle is a good idea, to keep the wood from slipping and eventually splitting. I like to punch a hole in both the hook band and the end band for a screw to help hold it in place. Put both bands on hot so they'll be tight when they cool, but don't burn the wood.

The handle gets a lot of strain, so it should be at least three inches in diameter at the hook band. For some reason, these handles are prohibitively expensive in hardware stores, although symmetrical and presumably easier to make than axe handles and such. I usually drawknife three-inch-square shagbark hickory for cant hook handles, sawn at the local sawmill and seasoned for several years. Or I'll split a three-foot length of hickory into quarters, rough shape with a sharp hatchet, then drawknife the finished handle.

A finished cant hook.

Tools to Buy

Most of your carpentry tools are better bought secondhand at sales and flea markets than made by hand. Handsaws and squares and folding rules are not expensive, as are the heavier tools for basic structural work. Your forge can supply you with all the tools the early settlers used, and these will let you build structures similar to theirs. For finished carpentry and masonry you could tackle making your own tools, but you'll approach a point of diminishing returns that's too significant to ignore. It all boils down to whether you want to use tools of your own creation or not, and how much that's worth to you.

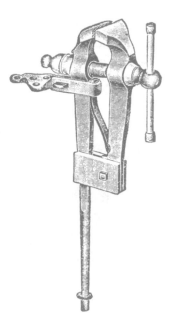

8

HOUSE HARDWARE

Blacksmithing for your homestead makes a lot of sense, when you consider that all the necessary ironwork for house and farm came from the smithy up until 175 years ago. There were cast-metal items from the foundry, too, but these were rare in the new settlements. Most of the hinges, latches, hooks, chain, nails, and even the tools were recycled from old horseshoes, wagon tires, and other worn iron.

Hand-forged hinges and latches complete a hand-made door. The hallmark for our log-house restorations is a custom-built door with hand-forged hardware.

Forging for House and Home

Certainly you can buy everything you need for a cabin, toolshed, and barn today, but let's carry on your preferred personal statement, the hand-built house, into hand-forged hardware. Look at your living room (perhaps your only room) — you can have front-door strap hinges and a latch of your own design; window hardware; the complete fireplace set of andirons, swinging crane, bellows, cooking grate, poker, shovel, tongs, and broom, with perhaps a stand for these; hammered wall sconces for candles or holders for kerosene lamps; ceiling lighting fixtures that may be for electricity or flame; coat hooks; and forged iron straps that support your ceiling joists from king posts above.

Wayne Haymes and I hammer a 36-inch strap hinge at his shop in Hollister, Missouri, using an 8-pound hammer.

And the kitchen. Here's a room that will keep your forge going for weeks. The pot rack, in any one of its interpretations, is a major project. We tried a circular hanging one, which held lots of pots, but it didn't fit with the shape of our kitchen. The present one is three successive bars mounted out from the wall with double hooks. The circular one, which is easier, I make regularly, sometimes with a heavy forged hook in the wall to hang it from. We'll discuss making this one later.

In forged household hardware, the principle of unity, or working from one piece of iron, is most important. Here your work is not just useful, it's also on constant display. A fireplace crane can be a vertical piece swiveling from eyes set into the masonry, attached to a horizontal piece with a hook, all braced with an angle piece. But it's more aesthetically pleasing if the entire swinging part is one piece, with perhaps the brace curved down to the wall and the hook integral. This could be welded, riveted, or joined with a loop-and-eye, but the one piece is stronger and more graceful, and (most important here) it looks better. We'll discuss forging one of these later, too.

I occasionally hammer a candleholder, complete with handle, from a single bar of iron. Sheet iron would be simpler, with the handle riveted in, but the grace of the single piece, and the evidence of more handwork, is pleasing.

Sliding closet door pull and bathroom-paper holder in our log house, both with hammered, veined-leaf design.

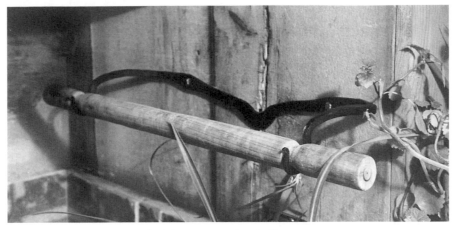
Forged paper-towel rack, holding a turned spindle.

In our kitchen, we use a lot of black cast-iron ware, and we hang them on individual one-piece hooks driven into overhead beams. Similar hooks with holes punched can be mounted with forged nails or screwed into the stud wall if you lack beams.

Utility First

My own approach to house hardware is utilitarian. Ornamentation is usually whatever I've seen and adapted, or a simple twist or curl that seems logical. I prefer to make hinges (for instance, for kitchen cabinets or the front door) as strap-and-pintle instead of double-strap. I notice this a lot in New England forgework, too. The pintle is simple, largely unseen, and strong, being a one-piece pin and mounting. Ornamentation goes into the strap. Of course, the original and still valid purpose of the strap of a hinge is to span and bind the several boards of the door. Long straps still give a look of strength, whether they're needed or not.

While we're on the subject, fake straps are a waste of time. The honesty of forged metal demands usefulness. We have a sliding closet door in our loft bathroom (no room to swing a door) with forged hardware. The pull handle is a simple loop, but it becomes a strap bracing the three 1-inch by 12-inch boards of the door. I had in mind a spade ornament at the end, but tried a leaf instead. Subsequently, forged leaves began to appear throughout the house, which seemed to fit with our near-wilderness location.

A couple we know live really far back in the woods in a frame house. They've undertaken to replace everything in it with handmade things, from leaded glass windows and copper lamps and handmade woodstoves to all the furniture and even kitchen utensils. I hammered a set of carbon steel carving knives for them in trade for copperwork.

That may be going a bit far for most of us, but certainly the feeling of self-sufficiency increases with the number of handmade pieces in use. Cup hooks, plant-hanging hooks, towel racks, mirror frames, and plain forged nails driven into log walls to hang things from — all are simple, useful items from your shop. The barn-board bed we sleep on is braced with forged pieces at the corners. So is the table we eat on. And shelves can be put up with those tin braces from the store, but your forged brackets will look better and hold more weight.

This is a Suffolk latch, with a one-piece forged handle.

Artful Iron

Essentially, any metal piece in your house can be made by hand if you have the skill, material, and time. The obvious dividing line for you is the time investment you want to make for the sake of forged hardware. I use a lot of forged nails and brackets, but I buy the screws. If something looks good and works with rivets, I don't buy bolts for it unless I'm pressed for time. I have made bolts, and sometimes still do, but except for rare uses, they can be had cheaper recycled or from the hardware store.

In our own house, I began replacing hinges and other pieces we'd bought in the hurry of building with forged work. Once the new pieces began to take shape under my hammer, I could see design possibilities, and I let the work grow asit would.

A fireplace poker being forged by my daughter Chelsea, then 11 years old.

Chelsea's completed fireplace poker.

The geologist's hammer I forged for Chelsea.

I am always impressed by the craftsmanship of a really great ornamental ironsmith, such as the late Samuel Yellin, and some of the young artists in Dona Meilach's book *Decorative and Sculptural Ironwork*. Few of us homestead smiths will attain that level of excellence, and that's not the direction my own work takes. I will always rather forge a drawknife or a mill pick than a chandelier, but I will do simple ornamental work, letting the decoration follow the function of the piece.

Making a Fireplace Crane

Your own house may become filled with artful iron, and I hope it does, but let it all have a use. Utility is most important, but there is room for art in useful iron. We mentioned the fireplace crane earlier, so let's go into some detail on this project. We'll do it in one piece. The size is up to you, depending, of course, on the size of the fireplace. It's a good idea to anchor the crane in the masonry of the sidewall, making it long enough to reach to the center of the fire. By pivoting the crane from near the front of the sidewall, you can swing it back pretty much out of the way, and bring it out away from the fire for fine touches during cooking.

Say we have a small crane, for a typical modern fireplace: an 18-inch crane, 12 inches high. We'll use a piece of ½-inch by 2-inch bar stock, about 24 inches long. Heat and reheat as necessary. First, bend a 90-degree corner in it about 8 inches from one end (the bottom) over the edge of the anvil. It doesn't matter whether you square the outside of the bend, but it will be neater if you do. So upset toward the corner as you bend.

Now, use the hot cut three-quarters of the way from the corner to make a cut down from the top to halfway through the horizontal, long leg of the bar. Turn the hot cut at right angles and split the bar down the center to maybe four inches from the end. Bend this cut bar out, and draw it from its nearly 12 inches length to about 18 inches long, smoothing the cut edge as you do.

Smooth the edge of the other half of the horizontal bar, too, but don't draw it. Now twist this bar so that the longer, thinner half is under it, and hammer the twist out. This will lengthen the bar a bit, but not much.

Now draw the end where both bars are still joined out to a long point. Bend it down into a hook for the pots to hang on, with maybe a little reverse lip at the tip.

Fireplace Crane

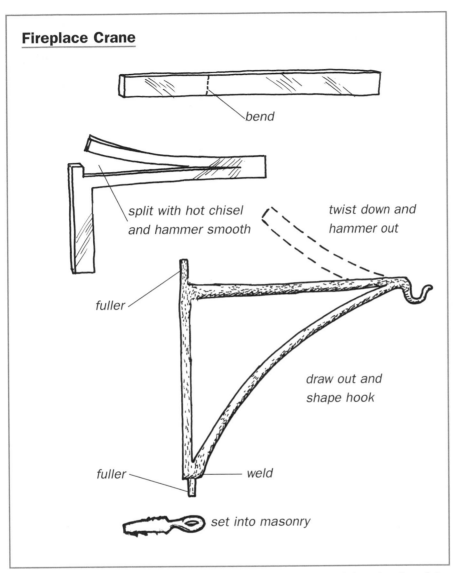

bend

split with hot chisel
and hammer smooth

twist down and
hammer out

fuller

draw out and
shape hook

fuller — weld

set into masonry

The vertical leg can now be drawn to near its 12-inch height. Scarf the end of both this and of the long, thin bar you bend down. Curve this bar to come against the vertical with a bit of pressure to hold it as you weld it. If the curved bar is too long, cut some off; if it's too short, draw it some more.

Heat, flux, heat, and weld this joint. You'll probably bend the brace bar in the process, but you can get the curve back later. Now fuller or hammer

Making a Fireplace Poker

1. One of my favorite fireplace poker ends is begun by drawing out a piece of ⅜-inch or ½-inch round stock, then bending it over.
2. The stock is bent, so that the point is laid back onto the shank. Then the joint is fluxed, heated, and forge welded.
3. The final point and barb are hammered. The other end may be a loop, basket, or knob.

a narrow neck in the end of the welded joint for a rounded protrusion one inch long with a substantial shoulder. If you don't have swages, just roll the fullered end as you hammer at the edge of the anvil. The shoulder will be for the mounting eye. Round the corner you left sticking up on top, too, for the other eye, but draw this one out about ¾ inch farther.

Now hammer loops in two square or round rods of ½-inch diameter, 6 inches or so long. Bend a 90-degree angle in each near the other end to hold the eye in the masonry. You can weld the loops if you like, but it isn't necessary; you can probably get a better circle if you don't. You can nick the shanks of these loops with the hot cut so they'll anchor better in the masonry, or split and spread the ends, or weld a crossbar onto each.

Now, if your fireplace is already built, we'll have to dig out some mortar between the stones or bricks to set the eyes in place. Figure your height to come out right beforehand, since it's harder to drill stone or brick than the mortar between. Dig out enough to allow the eyes at least three inches of depth, then push them in and mortar again, keeping it damp for two days to cure.

Now slide the longer upper pintle of the crane up through the eye far enough to let the bottom one clear and drop back in. There should still be plenty of the top pintle to hold. If there isn't, you can work on both the top and the bottom ones to get them right.

You may not be able to get a clean cut at the end where the two bars separate. By using the hot cut flat on the anvil you can cut almost to the end faster, but you get a tapered cut. Switch to another technique for the last ½ inch or so. Heat, then clamp the iron vertically in the vise and drive the chisel down from the end for a cleaner end cut.

If there are still rough places when you're finished, file them. I always heat again to black heat and quench in oil to get rid of shiny file marks. As a last step, unless I plan to paint the piece, I give it all the blacksmith's oil finish at black heat. You won't want to paint the crane, because the fire will burn it all off.

Of course, it would be easier to start with three bars of iron and weld up the crane than to do it from one piece. It would also be easier to arc weld the pieces, but we're talking about blacksmithing here, not the cold work you can find in any hardware store selling ornamental iron.

1. A pair of andirons begins with splitting a recycled bar of ½-inch by 2-inch mild steel with the hot chisel. Each step is duplicated with the matching piece, for better comparison and closer similarity.
2. One bar is spread over the anvil horn while the other heats in the forge.
3. The rough edges of the cut are smoothed, and the feet are formed.

4. The feet are flattened and finished. The hand-forged hammer used is my favorite, a 3-pound, 45-degree peen.
5. The stem of the upright half of the andiron is tapered, leaving metal near the Y for the horizontal half and at the end for the ornamental ball.

6. Metal at the end of the upright is folded over in preparation for forge welding (called a faggot weld) to thicken it for shaping a ball.
7. Borax flux is sprinkled on the joint at low red heat.
8. The faggot weld is made at white heat, with a minimum of pounding, to retain thickness. Sparks are droplets of molten metal and bits of scale.

9. While still hot, the welded end is shaped into a crude, elongated ball.

10. The ball is finished with a light hammer.

11. The upright is heated and punched with a square punch. A hot punch with a handle could be used here, to keep your hands away from the hot metal.

12. The hole, started from the other side, is finished by turning over the piece and punching through first the pritchel hole, then the hardy hole, as it is spread larger.

13. The horizontal bar of the andiron is shaped at the end into a tenon, to fit the punched hole in the upright.
14. Final shaping of the tenon is done with a file for an exact fit.
15. The heated tenon end is clamped into the vise.

16. The two halves of the andiron are riveted together with the ball peen. As the joint cools, it will loosen with contraction, so both pieces may be heated, or the riveted joint can be reheated and tightened.

17. The riveted joint.

18. The complete andirons. This set was a custom-made gift for the Virgil Culler family of Bethel, Missouri. It was for the fireplace I built them, in the hewn-log house they built with my assistance.

Making Hinges

We've mentioned hinges, and you'll need a bit of practice to get them right. I don't like to try to copy commercial hardware, with two or three loops around the pin. I form a single loop around a pintle, gate-style. When figuring the length of iron to allow for the loop, there's a bit more to it than pi (or 3.14) times the diameter of the pin. What happens when you make the loop is that the outside metal stretches, as you'd expect, but the inside of the loop upsets, leaving you with too little metal. There are several formulas different smiths I know use. One of these formulas works fairly well — pi times pin diameter plus half the pin diameter again.

I will admit to never measuring iron for a loop. I just hammer whatever iron I need to get around a forming pin of the right size, then draw the hinge strap to the right length. If the hinge is to be a matched, strap-and-strap one, I loop first, then cut out what I don't want with a hacksaw. That's for

I forged the warming-oven door and hinges shown here, as well as the iron-work reproduction chandelier, for an historic log house restoration in Virginia. The wood turnings and electrical works of the chandelier were made by master craftsman John Beard of Free Union, Virginia; he was also on the restoration crew.

a one-and-one loop, a two-and-one loop, or a two-and-two loop hinge with both straps even.

Drawing and tapering the strap, which is the usual treatment, will give you thicker iron near the point, unless you keep hammering it thin. This stretches the iron a lot, so you can easily get half again the length of the bar you started with. If you cut into the bar for ornamentation along the strap, you can get a taper of sorts with little drawing.

The ornamentation at the end of the hinge strap is as hard a project as you want to make it. A simple flattened circle is common and is a basic example of uncluttered design. These were usually done from just the metal at the end of the strap, and so were quite thin, stepping down from the strap thickness. I usually upset, or leave more iron when I taper, so I can do the circle full thickness. Form the circle while the iron is still thick, or it'll fold too easily. Then flatten after you have the basic circle.

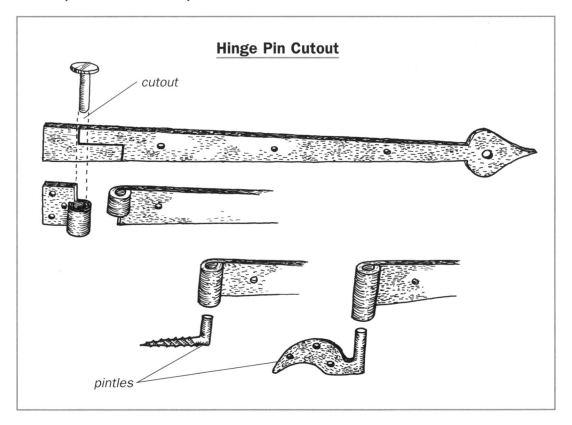

Hinge Pin Cutout

cutout

pintles

A Large Strap Hinge

Wayne Haymes at work on a large strap hinge. Wayne did commercial blacksmithing in Hollister, Missouri.

The hinge pin being riveted with the ball peen hammer for a decorative effect.

Artist Chandis Ingenthron, who has worked with me on several books, gets into the smithy now and then. Here, he has forged a double wall hook, with fleur-de-lis ornamentation. The flat surfaces of the hook were decorated with a design done cold with the center punch.

The spade is just a pointed circle. Both these should have a clean shoulder, done on the edge of the anvil or with a shouldering tool. This tool can be just a blunt cold chisel. Or you can use the edge of the anvil and a set hammer to get the sharp angle. Do the point of the spade on the horn for the reverse curve.

The double curve and the fleur-de-lis are accomplished by splitting the iron with the hot cut, then bending the points to one side for room to work them. This is close work, and like most ornamental ironwork, it takes patience and a light touch. I get a good idea of a student's control on a pair of fleur-de-lis strap hinges. Be prepared for almost instant heating of this delicate iron in the forge. It's easy to burn it, and the heat goes away quickly from these small masses of metal, too. Much of the final shaping can actually be done cold if you're careful and haven't picked up carbon steel by mistake.

When making hinge sets, it's a good idea to alternate operations between the two so you can keep them matched better. It's hard to remember how much iron you cut for a curl after it's drawn and finished. By doing one step at a time on each of the hinges, you'll get a better-looking set.

If you rivet halves of a strap-and-strap hinge, it'll almost always be too tight, like the tongs you made. Heat the joint to a medium red, then work the hinge until it cools. And oil it, or it'll tighten up again.

Your household ironware should have a finish of character. It doesn't matter how a cold chisel looks if it cuts iron, but you'll want the iron that's on view to look right. Don't try for that fake rippled effect that's supposed to look like hammering. A rule of mine is that every bit of the surface of a piece of ornamental iron should be hammered. And I don't try for a particularly smooth or a particularly rough finish, I just let it work out. It looks hand forged because it is.

Making a Pot Rack

A pot rack of some sort is almost a kitchen necessity, and I seem to make pot racks often. The most accessible is a circular one, to be hung from overhead or suspended from a heavy hook out from the wall.

I use ¼-inch by 2-inch stock for a hoop about 24 inches in diameter. A smaller rack of 16 inches or so needs to be made from 1- or 1½-inch strap.

Figure the length as simple pi times the diameter. Heat and bend on the anvil horn to form the hoop, correcting the arc as needed. Scarf the ends, and spring them together for a firm fit when welding. Flux when red hot, then bring to welding heat. Have the anvil horn close to make the weld on.

Now, you'll want three holes in the hoop, so you can strengthen your weld joint by punching one through its lap if you like (you'll be riveting a spoke through here). I use ¼-inch or ⅜-inch rod for spokes in the hoop, so punch the three holes the right size at equal distances around. You'll distort the circle somewhat by doing this, but it's easy to reshape.

Bunch the three spokes, with one extending about an inch longer, past the joint to be welded. Wrap the bundle with wire to hold it together. Now flux, heat, and weld, turning the bundle rapidly as you hammer it. Shape the weld, then draw the extra inch of rod and form a hook.

Circular Pot Rack

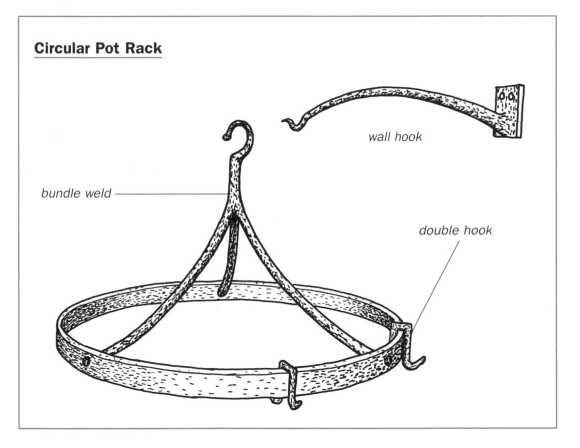

wall hook

bundle weld

double hook

Pot rack to hang over a kitchen butcher block. The square is ½-inch by 3-inch steel, riveted at the corners to angle iron for the loop-and-bar chain. The pot-hooks are single, split double, and bundle-welded triple.

A Basket Twist

The basket twist is made by forge welding several rods together at each end. Then the bundle is heated in the center and twisted in the vise.

Reversing the twist in the bundle-welded piece makes each rod stand out.

Three examples of
decorative basket twists
forged by Steve Stokes.

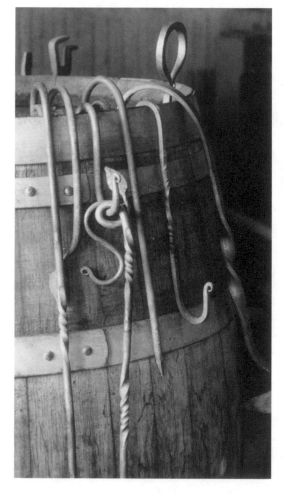

Ornamental iron forged by Steve Stokes,
master smith for Stokes of England in
Charlottesville, Virginia.

Working Cold

Veining leaves can be done cold, as shown. Then, when heated and curled, the veining curls with the shape of the leaf.

Riveting the square tenon of a hook for a paper-towel spindle into the leaf-design mounting bracket. Such riveting can be done cold for a tighter joint, or tightened later, since the heated tenon contracts when cold.

The length of these spokes is up to you, depending on how low you want the hoop to hang. I use a length 1½ times the hoop radius (one will be that 1 inch longer) and curve each spoke upward to the weld. You can curve the spokes before welding, but it's hard to hold them in line if you do. Shape them by eye, or draw out a curve on paper to match them against.

Put each spoke end through its hole in the hoop, heat, and rivet, with the spoke held in the vise. If the whole thing is lopsided when you're through, you can heat and bend one way or another to straighten it.

The pothooks I usually make are double and are hung over the hoop, as many as seems right. I use ¼-inch rod about 5 inches long, drawing both ends to points. Then I bend in the center, making a long staple. Heat the ends, and start both the hook bends at the same time on the anvil horn. You can work the hooks for a tight fit on the hoop or leave them loose. Another double hook can be made by splitting half the length of a ⅜- or ½-inch-square bar, drawing out and pointing each half for a hook, bent outward from the hoop and from each other.

And you can file notches in the hoop if you want the hooks to stay in place. Or even punch more holes and put the hooks through. My wife likes to be able to move the hooks for more or less room, so I just leave them mobile.

The late Rodney Harris of Montebello, Virginia, a master gun-smith, made this black-powder rifle boring machine from cherry wood with forged bracing and mechanism.

Another leaf-design hinge and the ¼-inch by 1-inch strap both halves were made of.

I got the idea for this pot rack from a unique wheelbarrow wheel a man once showed me. He made a riddle of it, asking whether I'd ever seen a functioning wheel without an axle. Or even a spindle. Well, no, I guessed I hadn't. So he produced this hoop, with three alternating spokes on each side, curved out from the rim, welded and shaped into two shafts. A neat job, and one you may want to try.

It simply means six spokes, with the odd ones joining at one side or the other. Care must be taken to get the wheel balanced, or it'll wobble easily. I would definitely rivet one spoke through the lap weld, since trundling the wheelbarrow around over rocks might eventually break it apart here.

Utility Meets Design

The number of household goodies you can forge is limitless. I like to make as much of the house hardware as possible, coordinating the design. So spade hinges might repeat a spade mounting bracket on a towel rack. Or spade ends on door handles and drawer pulls.

Our own design is a leaf of indeterminate species, which appears throughout the house. Hammering the leaf is a simple job, but it can be made fancy. The basic shape is just an elongated spade. The veins of the leaf are cut with a chisel while the leaf is straight (the veins are actually carved on the surface); then if you want a curl, the straight lines curl along with the leaf when you bend it. You could serrate the edges of the leaf with a three-cornered file for an elm-leaf look, but that'd leave you with sharp edges, which you might not want if you're easily injured.

Advanced ornamental ironwork, such as grape clusters with leaves and tendrils, will require a lot of practice and take a lot of time to execute. You'll use all the techniques of ironworking to produce the intricate detail of this kind of design. If this is to become your specialty, try to find another smith who does decorative work, and learn from him.

Remember a basic principle in your houseware forging: Function dictates the general appearance of the piece. A door handle may be as ornamental as a piece of Baroque furniture, but the handle must still open that door. I favor primitive blacksmithing, which is practically all function.

While a blacksmith shop isn't ordinarily the place for children, here three-year-old Amanda holds a piece to be worked cold, the veining of wrought-iron leaves.

9

HOMESTEAD TOOLS
AND REPAIRS

Your specific pursuits around your farmstead will determine how much your blacksmithing will help things along. If you plan a nearly self-sufficient operation, the uses of the forge are limitless. I have forged braces for a water wheel, brackets for a windmill, pieces for a sawmill, and may one day build a steam engine. The forge is a lot of help in the power-independence game. Certain needs and activities are common to most of us backwoods dwellers, and we'll talk about a few here.

A great deal of your time in country living will be spent moving things around — large things like stones, logs, lumber, firewood, gravel, sand, and other building materials. A pickup truck is handy; if you live in rough country a four-wheel drive is better. But there are jobs for which this workhorse just isn't suited, such as moving long logs or boards, or hauling stones too heavy to lift, or working in terrain too rough to build a road in. And you may not be able to justify the cost, either.

Two conveyances have been around farms for centuries, and they're still indispensable — the sled (sledge) and the cart or two-wheeled trailer. Both can be fashioned with the help of your forge and some wood, and with an axle and wheels for the trailer.

A cast-off aluminum pulley becomes a sheave for use on a lifting boom.

The Sled

Let's look at the sled first. It can be as little as six inches off the ground and still clear most obstacles, and its low profile means you can tumble stones, firewood blocks, or logs onto it without lifting. Its oak runners are quickly shod with two automotive steel spring leaves, and a floor of one- or two-inch oak laid across the top. Hooks or loops for the pull chain should be bolted to the runners themselves, since that's where the resistance comes from.

Heat the leaf spring runner steel and hammer it flat, then give it a curve upward in front to fit your shaped runners. Now heat the steel, and punch and drift out at least three tapered holes with a large center punch. This taper is to countersink big flat-head screws that hold the steel in place. Use the tie-bolt hole in the leaf for one of the holes.

The sled will last longer if, after you deck it, you turn it upside down and bolt in angle braces fore and aft to hold the runner to the deck.

Farm Sled

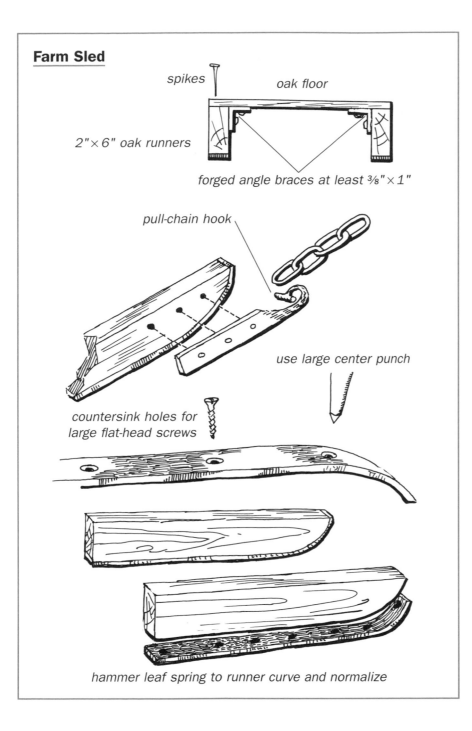

spikes

oak floor

2"×6" oak runners

forged angle braces at least ⅜"×1"

pull-chain hook

use large center punch

countersink holes for large flat-head screws

hammer leaf spring to runner curve and normalize

Some ⅜-inch by 2-inch steel is heavy enough, each piece about 5 inches long, and it'll keep the runners vertical. They get loose otherwise as you skid around rocks and stumps, pulled by mule, tractor, four-wheel drive, or your own muscle.

One bit of caution in using a steel-runner sled — it can run into, over, or under whatever is pulling it if you're headed downhill with a load. A pull device similar to an automotive tow bar would take care of the problem with a tractor or vehicle; your mule would need shafts and a complete set of buggy or wagon harnesses.

I have an old set of sled pull-hooks I picked up, forged from a wagon tire, that are designed to keep the chain from coming loose. The hooks curve back a long way, with a bend inward at the tip. The crossbar hook common in New England is another good idea here. The old-timers never had enough chain just to leave the piece on the sled when not in use; it was forever being borrowed for some other use.

Trailer and Hitch

The two-wheeled trailer is somewhat more complicated. A farm cart could have just about anything for wheels and bearings, but you'll use yours for highway travel, so plan to use an automotive axle, wheels, and springs. I've put together a few with no springs, but jarring under a load will eventually crack or break important parts when you need the trailer most — like when you have a load of sacked cement with rain approaching and no tarpaulin.

I prefer a conventional junkyard I-beam front axle from a pickup truck or pre–coil-spring car. A rear axle is heavier, which is sometimes better, depending on what you plan to haul. The dropped front axle usually allows a lower trailer. With it, I forge a clamp to hold the tie-rod, keeping the wheels straight ahead. You can also weld them straight, but that makes it harder to recycle the axle later.

If you have steel for framing, fine. Heavy angle iron, at least 4 inches wide and ⅜ inch thick, or channel iron is good. If these are not available, a good frame is 2-inch by 8-inch oak timbers or even 2×6s. I'm helping my neighbor build one with such a frame. I forged a set of spring shackles, the flat bars that hold a pin through the leaf spring end loops, and put ½-inch holes in them. Then we bored larger holes in the wood, lined with ½-inch water pipe bushings for the swing shackle bolts. The stationary shackles were bolted solidly to the wood. You may be able to find an axle with U-bolts, springs, and shackles attached. Unfortunately, later-model cars and pickup trucks don't have straight front axles; look for old stuff.

We reinforced the two-inch framing at the corners with heavy angle iron, which I forged. It was four inches high and bolted with four bolts at each corner for stability. Another cross member was angle iron braced across the center to support the tongue, which was also angle braced.

Diagonal bracing helped reinforce the tapered, 2×6 oak tongue. An unbraced wooden trailer tongue can snap off if you jackknife it while backing up or if you accidentally drop your anvil on it.

Decking this trailer with one-inch oak was simple, since it was all done between the wheels. My own farm trailer, with heavier axle, frame, wheels,

Forged farm trailer braces, at the hinged tailgate. Harness rings were added at the tops of the braces to tie things to. The wood was painted with oil sealant and red barn paint, the iron with flat black paint.

Making a Trailer Hitch

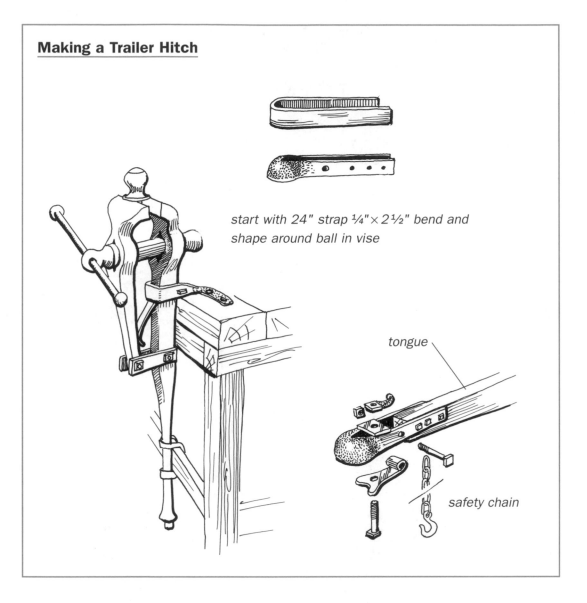

start with 24" strap ¼" × 2½" bend and
shape around ball in vise

tongue

safety chain

and springs, extends out to enclose the wheels in boxes. Forged braces hold
on sides as needed, and my tailgate and front wall are mounted on hinges
for extra length when dropped.

Now, I have often been in a location, or in a state of penury, that pre-
cluded a store-bought trailer hitch and ball. My Land Rover uses a heavy

This sturdy trailer hitch reaches far up under today's tin cars to attach to heavier framework. It's made of heavy angle and bar stock, forged, and arc welded.

pin, so my current trailer has no ball hitch, but I've forged several, and if you want to tackle it, I'll tell you how.

First, you need a ball. When I get a blacksmithing student who needs a challenge, I have him hammer out a trailer ball from one piece of truck axle shaft. It takes a lot of time, so you'll want to buy one, I'm sure. To fit the hitch, I clamp the ball in the vise and wrap a 24-inch piece of 2½-inch by ¼-inch mild steel around it, cupping it front and sides on top, and at the front on the bottom. Work it till it will drop over the ball nicely, then fit snugly as you pull the bottom lip back under it.

Now punch holes for mounting on the tongue, one for a safety chain and another pair of pivot bolt holes for the fork that will hold everything in place. This fork should be forged to fit the ball, with a loop around the pivot bolt. I make a square hole in the fork for a heavy carriage bolt to draw it up into place. This bolt goes up through a ¼-inch plate that can extend from the tongue as a separate piece or be welded to the hitch. I use a nut with a loop in it for safety wire so the nut won't unscrew until I'm ready for it to. Or you can thread a hole punched in a bar shaped to the purpose, which will give you a nut with a tail.

A Two-Wheeled Farm Cart

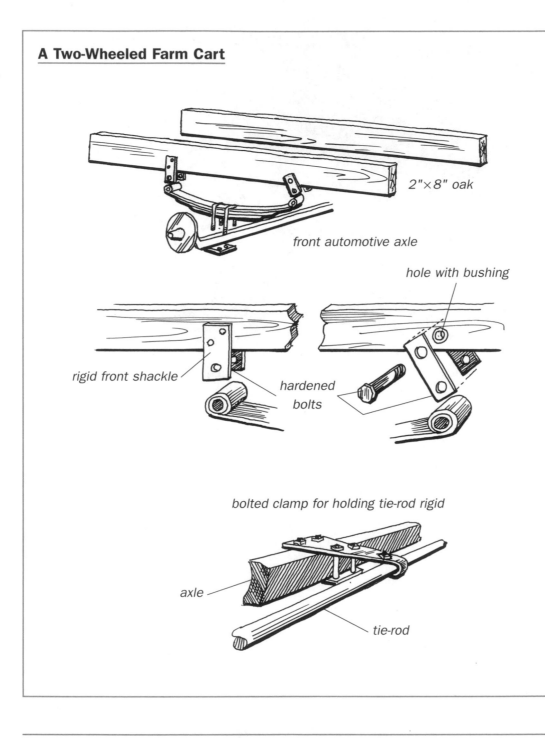

2"×8" oak

front automotive axle

hole with bushing

rigid front shackle

hardened bolts

bolted clamp for holding tie-rod rigid

axle

tie-rod

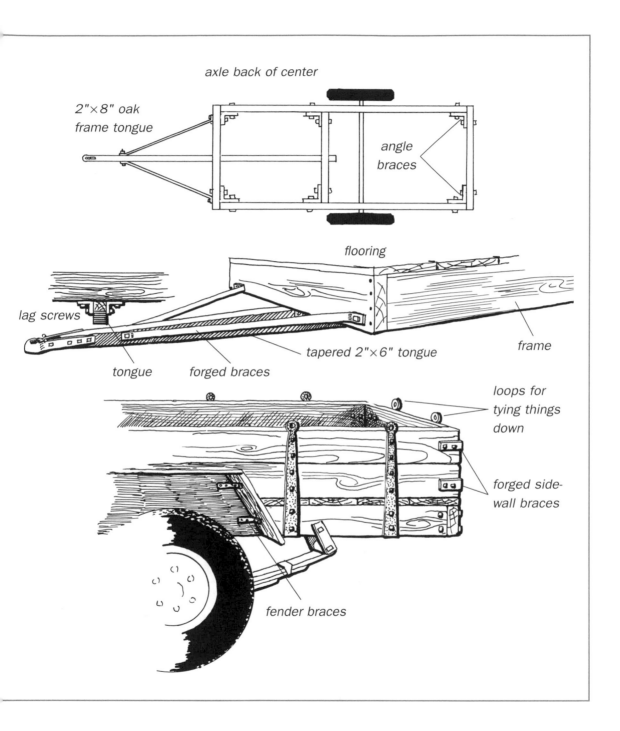

axle back of center

2"×8" oak
frame tongue

angle
braces

flooring

lag screws

tongue

forged braces

tapered 2"×6" tongue

frame

loops for
tying things
down

forged side-
wall braces

fender braces

Several of these homemade hitches are running around the country on trailers I've made, and they've given good service. As I said, though, you'd save by buying one already made and putting your expertise to work elsewhere.

If you put permanent sides on your trailer, do reinforce the corners with bolted angle braces. The vibration from country roads (*non*-roads, really) will soon loosen the tightest screw or nail, even in tough wood. And be sure you forge some loops and hooks along the sides for tying things in place.

I like to paint the ironwork black on a wooden trailer, no matter what color the wood is. And do keep your trailer or sled inside when not in use. In my part of the country, most of these useful contrivances rust or rot away in a few years when they could last generations.

Your four-wheel drive or pickup truck will probably have a bumper hitch already, but you may want to fabricate one for your car. Most post-1960 cars are unit-body, which means they're made of tin and not much else, so a strong hitch is a problem. I bolt bracing for steel to whatever looks solid,

My 8-pound splitting maul, forged from a junkyard sledgehammer head. The handle is drawknifed of shagbark hickory and has lasted many seasons.

and use a broad washer or piece of plate around the bolt to distribute the stress over as much tin as possible. It's important to brace for up-and-down thrust as well as straight-forward pull and side-to-side pull.

I once fashioned a hitch for our Volvo station wagon. It meant reaching far under for something solid to tie to, past the exhaust pipe and gas tank. I gave the hitch longitudinal stability with two ¼-inch by 1½-inch bars, on edge, bolted to heavy ridges up near the rear axle. Vertical bracing came down from the three-bolt bumper mounting plate at each side. Heavy angle iron went across to mount the hitch on, which was a piece ½ inch by 2½ inches, extended back. Then ½-inch-square bars are angled from the ends of this piece to as near the bumper mount as possible to take much of the stress from the angle iron. It is a tough hitch.

Tools for Splitting Wood

Let's get on to the useful tools around your farm or woods retreat. A tool you'll be forever grateful for discovering is the splitting maul. And it's become quite expensive, with the return to wood heat. As my blacksmith friend Wayne Haymes said, "If you want simplicity in this life, you've got to be willing to pay for it."

Forging a Maul

Here's what I've done repeatedly for myself, friends, and customers — buy a chipped sledgehammer head of about eight pounds at a junkshop or sale (never pay more than a couple of dollars for the head), and forge a maul from it. It pays to have help here, though I've done it alone with the short-handled six-pound hammer I use. Beating on eight pounds of carbon steel with anything less is a waste. Watch for folding as you taper the edge. I put a bit of reverse curve in on the anvil horn and hammer the final edge to the desired sharpness (but don't make it too sharp). Quench and temper draw to a purple; you'll use the hammer face to drive wedges and don't want the face to chip. I've had my share of flying bits of steel from factory tools that were left too hard.

Don't try punching a handle hole in a three-inch block of steel to forge a maul from stock. It's just too much work. If you make the maul or hammer this way, start your hole small and work up.

Wedges

Splitting wedges are easy to forge from large-diameter shafting of two or three inches. Try the finished product for hardness after quenching; you shouldn't need to draw the temper, since most shafting is medium carbon. If you do, temper draw from the hammer end. When this end mushrooms with use, heat and forge back into shape, or trim off the overhang. Again, a misplaced blow could send hardened steel flying if this is left too hard.

Cold Chisels

A cold chisel is among the first things I have a student make. Shearing rusty bolt heads, splitting rusty nuts, cutting bars and rivet heads — any cold cutting requires a good chisel. And a good chisel is one that's long enough that you don't hit your thumb or knuckle or skin your hand when the rusty steel you're cutting gives way. Too much live bark has been knocked off with those cute little toys you buy in their raised plastic-covered cards.

This toothed rock chisel is cherry red and ready to quench. Temper will be drawn to straw at the points. The teeth were ground and filed while the piece was annealed.

I use a bar of carbon steel, 100-point or so (coil spring or star drill shank), ¾ inch to 1 inch in diameter and at least 8 inches long. That much mass means you have to use a bigger hammer, but it's worth it to keep down the tetanus. Draw to a long taper, then shape the edge much more blunt, at a 60- to 90-degree angle. A keen edge will work on softer metals, but if it's hard enough to cut, it will eventually break or chip.

Sharpen while annealed, then heat and quench. Draw from the hammer end to a straw if spring steel, to purple if higher carbon. Experiment; you may want a harder or softer temper for your own uses.

Making a Cold Chisel • 1

1. First a section of automotive coil spring is heated in the forge and straightened in the vise.
2. An 8-inch piece is cut off on the hardy.
3. It's then straightened further.

Making a Cold Chisel • 2

4. Then the steel is heated and the tapered cutting edge hammered.

5. The struck end is tapered and trued up in the vise.

6. The chisel is annealed by cooling slowly from cherry red, in either ashes or sand, or, as shown here, next to the forge fire while other work is going on.

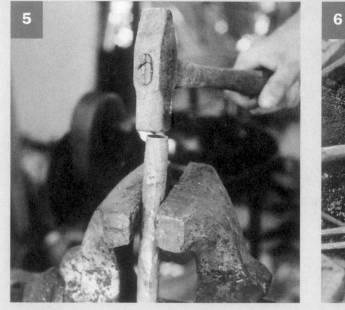

7. While softened by annealing, the chisel is ground to an edge, and any rough places are smoothed, cold.

8. The working part of the finished chisel is heated to cherry red and quenched to harden.

9. The chisel is honed or ground to a shiny surface so the tempering colors can be seen. Then it is laid on a low forge fire so that colors will run toward the cutting edge. This will leave the rest of the tool normalized, so as not to be brittle, but the edge hardest. A bronze at the edge is right for most work, using this 95-point steel.

For a better chance of getting a good forge weld, go through it cold a few times. Here, Wayne Haymes shows Bill Jennings how to hold the scarfed and lapped ends of a chain link for welding.

I usually taper the heads of punches, chisels, and wedges to lessen mushrooming. This will also keep the impact path nearer the center and lessen the chances of bending if you really get on the tool with a sledgehammer.

Hooks and Chains

For woods living, the cant hook we talked about earlier is a big help. I was raised around a sawmill and used one from early on. If you plan to do much with logs in their round state, have a hook or hooks that bite properly. And have some help around, too; about all you can say for the behavior of logs and other large masses is that they barely tolerate you.

Log grab hooks are good if properly made. Bad ones waste time as you back your tractor, mule, or four-wheel drive to rehook after they've pulled out. Two simple hooks on a ring won't do much. A scissors action is better, since the forward pull clamps the hooks tighter. But point angle and sharpness have a lot to do with whether grab hooks save or waste time. My best advice here is to experiment with your own pair if you plan to skid many logs. The hook is made the same way as the cant hook, but of heavier stock, about an inch thick.

Often I use a loop in a chain to skid logs, being mindful of the time I caught my thumb in one as the skidding tractor lurched forward. That was a dancing exhibition, with my peacock-blue thumbnail afire.

Making Chain

We talked about chain making earlier. And while it's not a profitable pastime, it's good to be able to link up stray pieces you might have found or traded for.

Round rod is cheap at the steelyard, or you might find some at your friendly junkyard. The early blacksmiths used worn-out horseshoes and wagon tires to make their chain rod, working up a supply when things were slow. You can reduce almost any scrap to, say, a ⅜-inch rod, which is a good size for strong chain. I use about a six- to seven-inch length for each link, scarfed and looped and linked, with no particular attention paid to final shape until after it's welded. There's plenty of heat left to shape the link after the weld is made and the joint shaped. Remember to make the weld at the end of the link; it's stronger there than at the sides, where machine-made

Making a Chain Grab Hook

1. A punched-hole chain grab hook. First, the bar is drawn out, fullered, and punched.

2. The half-bent hook is returned to the forge for a second heat. It would function this way as a slip hook, to let the chain slide in it.

3. The finished hook, with chain attached. Hooks must be heavier than their chain to keep from opening out in a pull. The grab hook holds across a chain link, not allowing it to slide.

chain is welded. Another thing I do is weld up, say, three feet of chain at a time, then join these pieces. It's easier than handling the entire length of chain every time you add one link.

Of course, here I am reminded of the old blacksmith's riddle of the man who brought five pieces of chain, of three links each, to the shop to have them joined, insisting that it could be done with only three cuts and three welds. The smith wanted to make, and charge him for, four. I'll leave that one for you to play with, and the answer makes my point about joining pieces.

Always keep short pieces of chain around to use with your ratchet cable hoist or winch. A good way to break a cable is to crimp it by hooking it around something and back to itself.

Very nice T-bar chain hook and swivel from New England. These welds were made on the anvil horn after splitting the iron, leaving heavier metal at the wear points. The T-bar was upset.

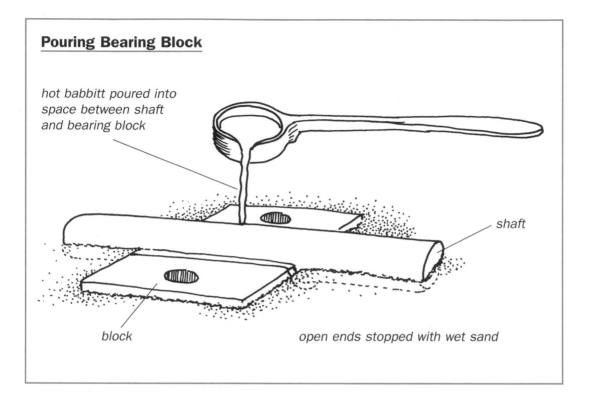

Pouring Bearing Block

hot babbitt poured into space between shaft and bearing block

shaft

block

open ends stopped with wet sand

Another story I like is of the old farmers who bought the first arc-welded chain when it began to replace forged chain. Suspicious of the new welding method, many of them would wrap the new chain around a stump they were burning or toss it into a burning brush pile to heat and anneal. They wanted to make darn sure any hard spots were softened before the chain was put to use. "Better to stretch than break" is sound advice still.

Forge your chain hooks from heavier stock than the chain itself. I once came by a bucket of old railroad spikes that made excellent hooks. Use at least ¾-inch bar for a ⅜-inch chain; otherwise, the pull will straighten out the hook before the chain itself gives way. Hammer the point and loop that links to the chain out while straight, leaving heavy metal at what will become the curve of the hook. Bend the loop and weld it before you shape the hook, in case you blow it. That's a good rule if your forge welding is a bit uncertain; weld early before you've put too much time into the piece.

Now shape the hook over the anvil horn, hammering flat the flaring inner curve as you go. For a grab hook, lay the curved hook on its back on the anvil and drive the hook almost closed, then open it from the tip on the horn. A grab hook should just slip over the chain thickness, with almost no clearance. Clamp a flat ⅜-inch-thick bar in the vise and do the final fitting around it. A slip hook should be deep, since it's easy for the chain to flop out, so use more metal.

I like the crossbar at the end of a harness trace chain or any pull-chain hook. To forge it, upset the tip first, so you'll have plenty of metal to work with, then work the bar as you taper to the point. Again, do everything on a hook before you bend it, or you'll have an awkward shape to deal with.

The "boomer," or chain binder, is a handy device that's going up in price fearsomely. I watch for broken ones I can buy cheap, then repair them. But if you forge this tool and leave a hard place in it through poor tempering, the jolts it will take can dump a load of logs or a loaded tractor off on somebody when it snaps.

A boomer should break over center just right and hold properly, so if you forge one from stock, I'd say have a borrowed one handy to refer to as you go.

In my growing-up days, we tightened chain around a load with a long pole through the slack, twisted around, and tied with baling wire. I've used the boomer to lift heavy objects, one chain link at a time, hooking my progress each time.

Do put a swivel next to each hook on the boomer. They're just about worthless without them. I picked up a good repaired one cheap once, because the smith who'd repaired it left out the swivels. He did do an interesting job on one hook, though, which had originally been the head of a pipe wrench.

I always riveted chain swivels to make the head, until I saw the New England style on some old chain I brought back from there. These swivels were forged by splitting the heavily headed rod; then it was put through the punched hole, shaped to chain thickness, spread, and lap welded on the anvil horn. This results in swivels much stronger than riveted ones. The riveted swivel also tends to spread and bind in the hole. The hole is punched in heavy stock, which is then drawn to chain thickness and lap welded over the head.

Old-Time Forging

Sharpening plow points used to be a standard job for country blacksmiths. It's a lot like drawing a blunt axe head. You're thinning the remaining metal to get a sharp point, then hardening and retempering. Often, you'll weld new carbon steel onto the old, as with the axe. It's a valid exercise for both axe and plow point if the replacement is too difficult, although as I mentioned earlier, a new point can often be forged just as quickly.

My brother and I used an old Oliver 20 sulky plow for years, and the flint rock on his farm dulled or broke several points. We were lucky to find points in junkshops and flea markets here and there for a while, but we in time used up the region's supply. I had resolved to forge the next one we needed from a grader blade cutting edge, but the occasion never came up. I doubt that we hit the last rock in his fields, but the plow reposes, replaced now, with a good point still on it.

The forge can be the saving grace of antique tools and equipment like that old plow. I have replaced many missing parts on my Model A pickup truck with hand-forged work, from spring shackles to bumpers. When your fondness runs to exotic or ancient machinery or both, your skills with iron will be tried. You don't find a crank or a drawbar brace for a 50-year-old tractor at the local parts store, so you make such a crank or brace. And the disk harrow you got free for hauling it away just needs a steel bar here, a brace there, new babbitt in the bearings, and it's ready for your rock patch.

I have been able to acquire good, solid machinery, from sorghum mills to cutoff saws and woodworking machines, cheap because the bearings were worn out. Pouring bearings is like most of blacksmithing — there's no magic at all, just a little horse sense. You need some babbitt, which most junkyards have, pryable out of rusting bearing blocks. And something to melt it in, which doesn't take much heat (I've done it over a fire of wood chips).

I pour the babbitt with the bearing block and shaft, if possible, set in place in damp sand. I use a cast-iron ladle heated in the forge. The stuff cools quickly, then I remove the block and repeat with the bearing cap, shaft again in place, ends stopped with sand. If the block and shaft can't be removed from the machinery easily, you can use a caulking made of clay, or buy some commercial waxy stuff the old-timers called "bear grease" to seal the ends, and pour while in place. Most older equipment bearings allow for shims, to be removed as they wear. I cut these from tin cans. And pour with the shims in place, to allow for more babbitt than you need, to ensure against wear. New oil holes can be drilled easily, last.

Making Knives

Knives find never-ending use around the farm. Hacking around your farmstead never seems finished — brush, briars, even tall weeds require a sturdy, keen blade to help keep your place from going totally back to the woods. The surplus-store machetes are usually made in someplace unpronounceable from old beer cans, or if they're good, are high-priced.

Spring steel, temper drawn to a purple or blue, makes a good machete (corn knife). Another good one is tough car-axle shaft, just quenched, if you'll take the time to hammer it flat. Or torch out a machete from old circle-saw blade. There are a variety of designs, including one with a sharpened hook on the back for close cropping without stooping. I use two, a heavy, long (24-inch) one for lopping limbs and small trees and cutting briars at arm's length, and a bowie knife, which is handy and much lighter. The bowie serves as every other kind of knife, too. I use it a lot to rough shape tool handles before I drawknife them.

Knifemaking is another venture altogether if you want to pick up some cash selling them. Most knife makers use expensive alloys and work cold, heat treating only after hours spent cutting, grinding, and belt sanding. Then they mirror polish them to compete for the high prices collectors will pay.

I am often asked to forge knives, and I explain that I am a blacksmith, not a knife maker. My concern is with forging steel, and I could care less about polishing out the forge and hammer marks. I want a blade of medium carbon that will cut, flex slightly without breaking, and hold a reasonable edge.

Folktales abound about knives that split bolts, stones, other knives, and hard objects in general and still hold a razor edge. And, of course, these magic knives will bend double and return to their shape an infinite number of times.

The facts are these — a lasting cutting edge has to be hard. The harder it is, the more brittle and more prone to chipping it is. The more given to flexing, the softer the metal. Of course, we now have sophisticated alloys for specific purposes, among them those of the stainless, no-rust character.

The magic folklore knife just doesn't exist. But really fine knives can be made from a variety of carbon steels or alloys; they just won't perform magic the way the aging town blowhard would have you believe. The best cutting edges, I still maintain, are to be had from carbon steel, soft enough not to chip, hard enough to last. The exact steel and tempering depend on the use the knife will get. A woodcarving knife is not a skinning knife is not a throwing knife.

I once watched a knifesmith impress the other rustics by shaving steel strips off a factory cold chisel with one of his knives. Of course, that

A vegetable chopping knife being shaped while annealed, and a finished bowie knife, both of spring steel.

particular knife, no doubt quenched hard, was not for sale, and probably would have shattered if dropped on the sidewalk. His other knives were acceptably flexible and, I'm sure, quite serviceable, but not for shaving chisels. I really don't know if he'd annealed that chisel or not. He probably had.

You can buy factory-made hunting knives for much less than the time investment you'll make in forging them or for the price you'd have to charge a customer. But here again is that perfectly valid mystique of the hand-forged tool that lets us blacksmiths survive. Nobody's grandchildren gaze in awed admiration at a knife that's still available downtown, cheap. Remember that each knife you make is unlike any other in existence; thank heaven for handcraftsmanship.

I like to hammer a knife to a state that requires little filing or grinding. The flatter is useful here, to leave a nearly dent-free surface. I have to know what a knife is to be used for before I start, and I select the steel and temper to that use. It may be evenly tempered, or hard at the edge and quite soft at the spine. I may sacrifice edge holding for flexibility, as with a throwing knife or machete that gets jolted a lot.

I use a steel anywhere from 30-point to 80-point carbon. Never have I tried to use an old file, which just isn't very good steel for a knife (it's dandy for a file, and was often the only steel the old-timers could get that they knew was high carbon). Knifesmiths mostly lean toward 4140 or D2 carbon or the 440 stainless group of alloys. These are medium carbon and have some nice characteristics. Remember, these last two numbers often denote carbon content.

I always quench a knife in oil to avoid cracking. Warping, too, is lessened with oil. If a knife warps, you can go ahead and draw the temper, then cold hammer the convex surface with it flat on the anvil to straighten it. Stresses in the steel are relieved this way, too, and if you've done your job right, you will not break the blade. If you've blown it, you will again, and will shatter the knife. So start over. Quenching straight down in the oil will usually forestall warping.

Knife making is a high craft form, so experiment around with steels and tempering before you call yourself a knifesmith. I'm perfectly content to avoid that label, though a lot of my knives are skinning creatures, carving wood, and hacking brush around the hills at this moment.

Another thing I'm not, I find, is a farmer. I just have no talent in that direction, although I spend a lot of time helping build or repairing implements for folks around me who are. If you dig in the soil, you'll keep your forge forever busy.

And as I said earlier, the extent of your work with your place in the country will bear a direct relation to the time you spend working iron.

So hammer your way toward the goal of independence, whether you ever actually arrive or not.

Wayne Haymes points out areas to be draw filed while this hammered knife blade is annealed. Farrier Jay Andrews and silversmith Ginny Ward were students in our blacksmithing workshop.

10

THE FARRIER'S CRAFT

I cannot tell you how many times I have been accused of not being a "real" blacksmith because I don't shoe horses. In the memories of the living generations, most country blacksmiths, driven by the machine age to accepting any and all kinds of work to survive, shod horses. And that picture has become fixed in the minds of just about everybody, except smiths and some farriers.

"Sure is a funny-looking horseshoe you're making," calls the craft-fair wag, popcorn in hand and mind left safely at home, to the blacksmith sweating over a trivet or scroll.

"You think this horseshoe looks funny, you should see the horse it goes on. He's a real mess." Rehearsed good humor, although the smith would rather say something else entirely.

What "Real" Blacksmiths Do

In pure fact, shoeing horses requires a knowledge of the animals and their physiology completely foreign to iron. And conversely, few working farriers know the mysteries of tempering steels and the wide range of other smithing practices.

But the myth persists. Those old-timers saw every country smith shoe horses, and you just ain't a blacksmith unless you do it too, Sonny. No good explaining that horse owners and professional farriers did most of the shoeing up until the general smith got into it too, maybe 150 years ago. Anyway, what does it matter?

"Well, the real reason I don't shoe horses," I'd confide to the craft-fair crowd, "is this deep-seated mistrust of horses." I'd prop my foot on the anvil block. "You see, one threw me once, years ago . . ."

Ah! Psychological stuff, right off the TV soaps; they're on home ground; they begin to nod sagely. ". . . and I landed on my face right in these sharp flint rocks. Well, it took out a whole big piece of my cheek, see, and I went to the doctor."

"'You're gonna have to have a skin graft, Mac,' he said."

"'Fine,' I said. 'Whose do we use? Maybe somebody without too much acne.'"

"'No, no, we have to use yours. Pick a spot where you'll miss it the least.'"

"Well, I thought about that some, and decided my rear end was the safest spot. So that's where he took it, and he patched up my face. And you can't hardly tell it now, can you?"

"Sure did a nice job," they say. "So that's why he don't like horses. Does he really mean all that? These flint rocks'll sure cut you . . ." nods, mutters.

"So now when my mother-in-law visits and kisses me on that cheek, I just smile . . ." and I can get back to work till the next crowd gathers.

In fact, being a neighborhood farrier is a good way to make a few dollars. There are an awful lot of horses in the country today; someone recently told me there are more now than in horse-and-buggy days. Maybe so. The folks who own horses usually own several, and there are lots more folks today. Someone has to shoe those horses, particularly if they're to be ridden, worked, or shown. Lots of people have theirs shod in any case to protect the hooves from excessive splitting and pasture-stone damage.

Advice from Pros and Homesteaders

So we'll talk about the farrier's craft, which could become a very real part of your homesteading if you elect to go into horse-powered pursuits. And for your own animals, my brother John, who's his rural neighborhood farrier as well as a farmer and a smith, has some good advice:

Before you buy any horse for any use, have his feet looked at carefully by a good farrier. No need to set yourself up for a continuing problem of corrective shoeing, unless you happen to be good at it. Too many prospective buyers ignore the feet, which may not show in gait if shod correctly.

Another bit of advice — get to know a farrier, and weasel some way to be his apprentice for a time if you're serious about this. You'll either give it up or learn it well. You can enroll in a farrier's school, too, of which there are several good ones across the country, but they aren't cheap. The schools will usually teach you hot shoeing and basic blacksmithing, but you already know that. Some farriers never heat a shoe to work it, but do it all cold.

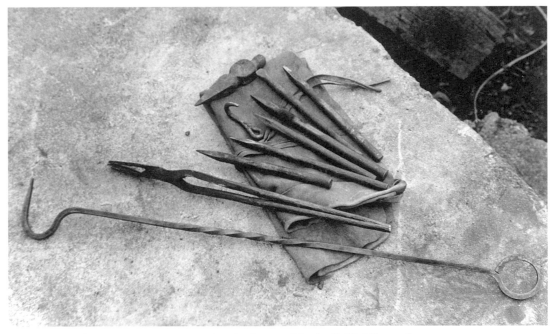

Some of the farrier's tools made by Jay Andrews of Springfield, Missouri, during a blacksmith course I taught.

Some working farrier tools. 1. clinch knife. **2.** clinch cutters (or nail cutters). **3.** hoof pick. **4.** hoof knife. **5.** hoof nippers. **6.** rasp. **7.** pritchel. **8.** shoeing hammer. **9.** alligator. **10.** 3-pound forging hammer. **11.** leather apron. **12.** factory and hand-forged shoes. **13.** ball-peen hammer.

Bending the point of a barred shoe cold is no good, though, so apply your forging knowledge to the teachings of a cold-shoe farrier.

A good shoer makes it look easy. Bob Peters of Berryville, Arkansas, could shoe 30 horses a day, partly because he was an experienced professional, partly because he was built close to the ground and a powerful, stocky man.

"Couldn't shoe 10 if I had to bend my back," he said. His hands moved with no apparent effort, and the hoof was pared, trimmed, and rasped, an apprentice often handing him tools and helping hold a restive horse. The shoe went on with a few lightning hammer blows, with those wicked nail points twisted off sort of in the same motion as the last stroke. The clinches were made and rasped smooth while Bob talked horses to his client and reassurances to the animal. A quick rasp around above the new shoe left it all smooth and fitted, and the hoof was let down.

Jay Andrews of Springfield, Missouri, was a young farrier, a graduate of a Midwest school. He took a course from me to learn more about smithing in general. His equipment was complete, down to a gas forge in his pickup bed. His approach was also professional, though somewhat more measured and methodical. Like all farriers, these men had a rapport with the animals that was instantly evident. Without that, you would never be a farrier.

Shoeing a Horse

I wanted to watch a homesteader at work, not a professional, so I went over to my brother John's place to see a typical farm shoeing. He was to put shoes on his Arabian mare, who had some registered name, but his teenage kids called her Dad's Root Beer, partly for her color, a foam white flecked with brown.

Well, Root Beer had borne a foal and had run free over a year unshod, down in the creek-bottom fields. So her hooves were chipped and split pretty deeply, but John said that's natural, and a couple of shoeings would even everything out again. His then 16-year-old daughter, Maria, wanted to work with the mare — a very fast horse — hence the shoeing.

John has an empathy with animals that once enabled him to calm a wounded deer, then carry it single-handedly out of the woods for doctoring. His horses are like faithful dogs with him, which makes shoeing easier.

First Root Beer was tied loosely to a tree, then John wiped some insect repellent on her legs to lessen a tendency to kick at flies. Then he picked rocks and bits of wood from her front hoof with a hoof pick that he had forged from a worn-out horseshoe. Next he explored the entire bottom surface of the foot with the hoof knife, paring here and there to remove dead horn and look for problem areas. Satisfied, he used the nippers to clip off excess growth of the outer shell of horn, which is more like trimming fingernails than anything else. He didn't try to cut to the depth of the

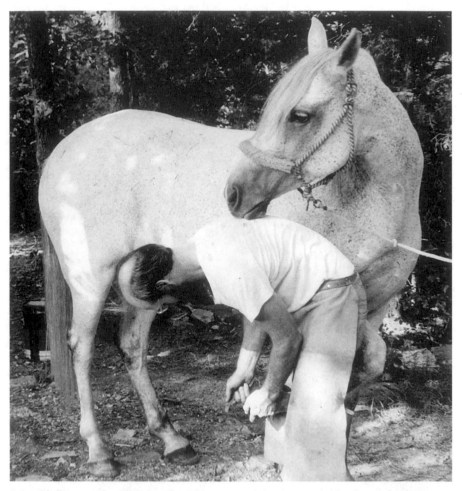

John McRaven prepares to shoe his mare. John is a neighborhood farrier at his farm in Hilda, Missouri.

broken-out places, which might get down to sensitive tissue. The bottom surface of the hoof was rasped flat, then a shoe of the right size tried. This mare took a #1 shoe, but it was obvious that it would need work.

John uses a homemade shoeing anvil on a heavy spike for shoeing, and except for corrective work or specialty shoes, he shapes them cold. They're low enough in carbon that they can be hardened only a little by quenching from red, so they aren't that difficult to shape cold. A barred or caulked shoe he would have worked hot.

The shoe is fitted so that the line of recessed nail holes falls just outside the white line of cartilaginous growth inside the outer horn. The nails are pre-bent to follow a path to the outside of the hoof, so of course there's a right way and a wrong way to drive them. Sometimes, too, the prepunched nail holes in the shoe will have some iron left in them, which can be cleaned up with the pritchel (square-hole punch).

The shoe is held in place with one nail started, then another is driven across the hoof from it. Hammer blows must be light and quick, since this is the part the horse enjoys least. The instant each nail is driven, the shoer is holding a deadly weapon in his hand. The pointed nail, sharper than a lot of knives, is curving out of the side of the hoof maybe an inch. I have seen fingers laid open, legs ripped, and even the tough leather apron of the shoer torn when the horse decided it wanted its foot back on the ground at this point, thank you.

John has developed an extra quick hammer stroke, switched to the nail point that bends the nail down almost parallel to the hoof. Horses obviously know when you're through hammering, and they often wait till then to jerk the foot away.

Now, even bent down, that projecting sharpie has to come off quickly. Some shoers use the clinch cutters to nip it, some slip the hammer claws around it and twist it off. The object is to get the hoof disarmed, fast. Subsequent nails are cut as driven. If the farrier hasn't let the horse put its foot down for a while, it's a good idea now. It gives the farrier a chance to straighten up, too.

Next, a small burr of cut horn fibers will have been pushed out by the emerging nail, and this should be cut off with the corner of a file somewhat finer than the rasp. A little horizontal groove is made just under the cut-off nail, so when the nail is clinched it won't tend to start a split in the hoof.

1. Stones are removed with the hoof pick.
2. The curved hoof knife pares away dead horn.
3. Hoof nippers are used to trim excess horn and even it out to take the shoe.

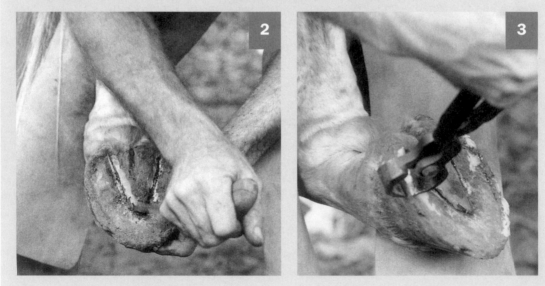

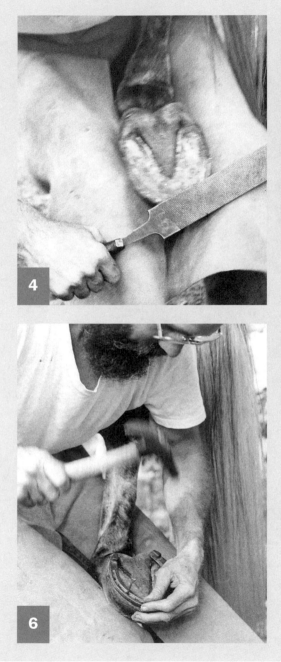

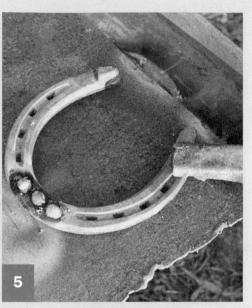

4. The bottom of the hoof is rasped for a flat surface.
5. John uses standard shoes, but he builds up three spots of hard-surface electric welding rod at the toe on his own horses.
6. The shoe is nailed with the nails slanted outward at just the right angle to emerge from the outside horn at the right height.

7. The sharp shank of the nail is either twisted off, as here, or cut with nail cutters as soon as it is driven.

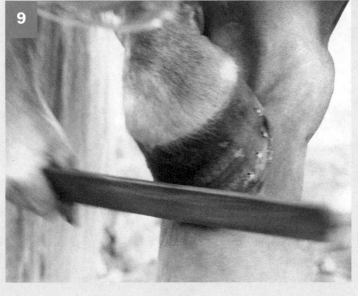

8. This tool, called an alligator, clinches the nail and shears off some of the ragged cut end. A horizontal groove has been cut just below with the file.

9. The nail ends are finished off with the rasp, and the hoof is smoothed even with the shoe all around to finish the job.

Some farriers use the clinch cutters held closed, butted up against the nail head, while they clinch the end with the hammer. Others use an alligator, which looks like tongs with one jaw bent out in a curve with sawtooth grooves across it. John made his from worn clinch cutters, but they can be had new. The flat jaw holds the nail head, while the bent-jaw serrations catch and clinch the cut end. A little more pressure and the hardened serrations slip, shaving the clinched end smoother. These ends should then be rasped, or the horse can hurt itself, hang up in fences, or hurt other animals, including you.

John gives his mare's hoof a final rasping to take off any overhanging horn, and the foot is shod. He likes to start with a front foot, then work around both back ones, finishing "with another easy one" at the front.

Had there been old shoes on, he would have used the clinch knife, driven flat between shoe and hoof to shear the old nails off, then cut off the old clinches and pulled the nails with the cutters. Or he might cut off the clinches and get the cutters under the branches of the shoe to pull it as he clamped.

Horses in flint rock country go through shoes fairly fast, so most folks put caulked shoes with toe bars on. John usually uses a hard-surfacing electrode to arc weld three humps at the toe for wear, since that's where it's heaviest.

"Makes it easy to track my horses if they get out, too," he says.

The list of specialty shoes that have or have had practical applications goes on and on. One old-timer once offered to show me 78 kinds. I got lost after a dozen or so. There are toe bars to be welded on, drive caulks, caulks made simply by bending the shoe's metal ends for more iron. Lots more.

Making Shoes

But as I said earlier, unless you've bought yourself some problems in the horse or in the location of your homestead, standard shoeing practices should do. You can, of course, make your own shoes from stock, using mild steel or low-carbon bar $5/16$ to $3/8$ inch thick, $3/4$ inch wide, and about 1 foot long for a start. I cut the nail head grooves while the bar is straight, using the hot chisel.

I understand farrier's suppliers have curved chisels for this. Use a store horseshoe for a pattern, and heat and bend to shape. Start at the center, or toe, and work to the sides, or branches. You'll find that front shoes need different shaping from back ones, so you might as well get them as nearly right as you can while you hammer them.

Use the pritchel to punch the nail holes, four to a side, and try to space them differently. One problem with machine-made shoes is that the nail always goes in right over the old nail hole, and hoof horn is weaker there. Angle the pritchel so that the nail hole will lead the nail outward through the hoof wall.

Round the ends of the shoes (after you've determined the length) by close hammering; or you can heat the ends of the shoe, clamp the shoe in a vise, and rasp the ends while still hot. Don't toss the shoe into the tub if it has any carbon content to speak of. You'll be bending it a bit here and there later, so keep it soft until it's the right shape. Be sure to let it cool somewhere away from the reach of small children.

Everyone knows the story of the smith who was plagued by the village smart aleck, a stripling who always got the last word. The smith had just completed a set of shoes for the farrier, and had left them on the anvil to cool. They looked innocent enough, at black heat, as the irritating fellow made his routine entrance.

"Now, don't you start," the smith held up his hand, "until I've got your advice on something. You bein' a man that gets around among the gentlefolk a lot, I want you to check over those shoes I made for the judge's mare, and tell me if you think they're good enough."

Well, entrusted with the function of Quality Control Supervisor (he could just see the name in Barnum script), the lad approached the anvil and, of course, picked up a shoe. Then with a howl he dropped it and plunged hand and arm into the slack tub, amid the roars from the smith and assorted gathered town worthies.

"What's the matter, Son, was that shoe still hot?" queried the smith, in a vain effort at solicitation.

"No," gritted the youth, then righting himself and adjusting his straw boater before marching out. "It just don't take a man like me long to look at a horseshoe."

For my own part, I doubt if I'll ever undertake shoeing horses, unless forced by grim economic necessity to run my life around them. If that happens, I'll have to get on a lot more familiar and trusting basis with horses first. That big an animal can cause a lot of damage just accidentally.

Besides shoeing, there are many other smithing aspects to the world of the horse. There's the harness, with buckles, rings, chains, and hooks. There is the need for ironwork in building, repairing, or maintaining everything the horse moves, from plow to sleigh. Man and horse had a long working relationship on the small homestead for many hundreds of years, and there are some quite good points for the arrangement.

One thing you will be able to count on is a lot of work hammering the iron you and the horse will wear out over the years. And shucks, then you'll be a "real" blacksmith, too.

11

INCOME
FROM YOUR FORGE

First of all, if you have any ideas of slipping into the role of the old-time blacksmith and earning a living by the sweat of your hammer like great-grandpa did, you can just about forget the whole thing.

The economics of competing with machine-made ironwork drove the old blacksmith out of business. First, he broadened the range of the jobs he did. I mentioned horse-shoeing earlier, and wheelwrighting was another common area. Lots of village smiths were also the local dentists, pulling teeth by hand or with tongs. Certainly this man who made the early dentist's tools knew something of their use. The smith made coffins, too, and occasionally tinware, and he expanded his smithy sometimes to make room for a foundry operation.

But it was a losing game. One by one the old shops closed, the ranks of thousands of smiths thinned by age, or by their going into new specialized fields — welders, machinists, and mechanics were often former smiths. So were many of the farriers, obliged by the changing times to become traveling craftsmen.

Finding Your Niche

A few smiths, such as the great Samuel Yellin of Philadelphia, attained an excellence in decorative work that exceeded anything the machine makers could produce, and they continued a thriving business. I have read that Yellin had as many as 300 smiths working for him at the height of his productivity in the second quarter of the 20th century.

Others specialized in gunsmithing, or in fine cutlery, or in one-of-a-kind tools of carbon steel for discriminating buyers. Indeed, these two forge products — decorative ironwork and hand-forged tools — seem to be about the only ones saleable at a profit today.

You just can't make it by sharpening plow points and making chain. The rare individual who wants forged chain will elevate his eyebrows at the dollar-plus per link you'd have to charge. And even the sign bracket you labored over, drawn and scrolled, riveted and graceful, can be had cheaper done cold and arc welded by the ornamental iron shop that turns them out by the thousands.

True, there are those few people who want and will pay for art in iron. But you must know you will be joining the ranks of artists and craftspeople starving all across the land today, competing for sales to these few. Potters, painters, sculptors, leather workers, furniture makers — talk to them about their customers, and about the lookers, whose houses are full of plastic and who can't tell the difference. And while you may be the only blacksmith at, say, a major crafts event, you'll face the same general indifference to hand-made work as these other creative people do. The public is so conditioned to machine-made duplication, most buyers really can't understand that your work is all done by you and that each piece is unique.

I'm reminded of my friend, sculptor Dennis Thompson, who's fascinated with dragons. Powerful, intricately sculptured dragons with hundreds of body scales attached separately, each porcelain tooth shaped and inserted

— art in a very high form. At an art show once, a slow day, nobody buying, a woman tapped one.

"Is that clay?" she asked.

"Yes ma'am, it is."

"Well, it's made in a mold, of course, then."

"No ma'am, I built it all by hand." Patiently. (How many thousand times had he said this to people?) First of all, no mold could enclose those shapes, and second, how would you get just the porcelain down into the claws and teeth, and then . . . oh, never mind. The woman was indignant.

"You can't expect me to believe that wasn't done in a mold, young man," and she turned to leave. .

"Well, all right, lady, I'll admit it was done in a mold, but that mold," he shook a finger at her, "that mold was a real son of a bitch to build!"

We have one of Dennis's dragons, but he refuses to let me see the fabled mold.

My commercial work has always consisted mainly of tools, mostly of an antique nature. Drawknives, froes, and broadaxes sell, as do some more present-day useful tools. I can always sell a digging bar or prybar of carbon steel with a chisel point. Or a tempered punch or cold chisel, if cheap enough. Buyers want to be able to brag about handwork and, as in the old days, will attribute almost magical qualities to the steel you've hammered, especially if they've watched you do it.

Where to Sell Your Work

At a crafts fair I can almost always sell a fireplace poker or pot rack I've made right there before the onlookers' eyes. Or, as often, I can take a custom order for one just a little different to suit a particular taste or need. In the matter of punches and chisels, showing the magic of hardening and tempering never fails to draw a throng of handymen. Then a scornful comparison of hand-forged tools made from selected carbon steel (with hardness graded to its use), against those machine-shaped, mass-quenched offerings of the hardware store, will generate some sales. During my soapbox lecture, I hint at the doubtful origins of factory-produced tools, and have been known to accuse those villains of recycling beer cans to make cold chisels. (Only now that they're all aluminum, it's not easy.)

The fair circuit is fun, but it's hard work. You need a portable setup, with a small anvil and a light forge. You'll get into some good fairs because your demonstrations will bring crowds, and the local media will zero in on your colorful setup. But be prepared for endless and dumb questions, and prepare to learn that everybody's grandfather was a smith. Only a real smith (better than you).

So many fairgoers are there to be entertained, not to buy. Many will buy trinkets or souvenirs, and you can make your entry fee back by selling things like hand-forged nails at 50 cents or pothooks for a dollar or two. You'll sell or trade pieces with the other exhibitors, who are appreciative of this kind of work and often delightful people. The camaraderie at a favorite annual fair is as big a reason for my participating as the prospect of sales.

Wholesale and Mass Production

Selling through other shops puts you into the wholesale production business, and handwork is hard to wholesale profitably. A good smith I know went this route and was broke in a year. The standard price around the Midwest is 50 percent of the retail, and shop owners will try to get you to lower your price more, so the pieces will move.

Consignment is an inhuman way for the craftsman to sell. Percentages to the retailers are high — 30 to 50 percent — and there's no money till the eventual sale. In the meantime, things get lost or shoplifted, and you have no money to buy iron with — let alone food.

You can go into mass production and cut prices and quality to compete for the vacationer's dollar. But of course, then you're no longer a blacksmith. You can also make each piece the only one of its kind — art in iron — and price it accordingly for that rare buyer who must be out there somewhere. And be assured that you have not compromised your art, as the utility bills and the rent come due and your stomach growls.

Selling Online

A bright spot today is the Internet, which allows you to advertise and sell your work directly to the consumer at retail prices. You can compete, because nobody's marking up your prices. Economists predict that this medium will allow the return of the small craftsman, small farmer, and small businessperson in general. I hope it does.

Marketing is why so many blacksmiths are only hobbyists today. Making a living at it means moving fast and leaving no stone unturned. The smiths I know who're not salaried by traditional arts centers do some of everything to make it, as the old-timers had to.

Finding an Outlet

I teach workshops; do a few crafts fairs; appear at school programs and civic clubs; write articles about blacksmithing; and make just about anything a customer wants, from current orders for a set of kitchen knives to a massive rake for grubbing flint rock. And I can't begin to make a living from my forge. If I placed pieces in all the crafts shops in the region, the law of averages would lose some for me, make some enemies for me, work me mercilessly, sell some for me, and I might make a sort of living at it — if I were single and living on a place that was paid for and didn't spend much money.

I do an odd job for a farmer now and then, like forge a digging bar or a hook he can't find in the hardware store. And welding-shop owners will bring me a problem they aren't sure about, usually having to do with cast iron or carbon steel. (Why does the spring steel keep breaking right next to my weld? Because the cold steel next to it is chilling it and making it brittle.)

A good connection is with one or more antique shops. Missing hardware is common, and I do pieces for old furniture, primitives, and collectibles I don't know the names of. Custom-furniture makers are also good people to work with — the very nature of their work and their clientele fit with your forgework.

Working the Craft-Fair Circuit

Traditional arts centers often hire smiths during tourist season. This is show-off work, mostly, and you get tired of answering incredible questions. But I did learn more than forge welding from the late Shad Heller, who was the star attraction for 20 years at the center near my Missouri home, and he knew his smithing.

Bob Patrick earned his living as a smith for many years. He taught almost constantly, working on his own and through the local arts council. And his ornamental work was among the best I've seen. He was located in Bethel, Missouri.

Wayne Haymes ran his shop in Hollister, Missouri, full-time, supplying fireplace sets, triangle gongs, and horseshoe puzzles to retailers in quantity. He also did custom work, from boot scrapers to belt buckles. But he eventually had to take a regular job, smithing then only on days off and as special orders required.

For several years, I did the craft-fair circuit as a blacksmith. I never made much money at it, but I made some good friends and enjoyed it, most of the time. Although most fairs are in the heat of the summer, and pounding red-hot iron out of a flaming coal forge can stop being fun.

But the fair organizers, from harassed, first-time schoolteachers in backwash communities to smooth professionals in charge of established moneymakers, were always glad for the color and clang of my setup. At any given time there'd be a crowd around as I shaped the glowing "black metal" to the steady patter of lore and jokes I'd stored away.

I hammered and welded link chain and made fireplace pokers, cold chisels, knives, and nails, mostly simple stuff that didn't require too much concentration. And I'd listen to all the old folks whose fathers or grandfathers had been smiths like me (only better, of course, as they remembered it, having been much older and wiser).

The other craftsmen were almost always a joy to visit with, camp with, trade with. There was Jim Lakiotis, a carver of wooden spoons, and his wife from West Virginia, with five daughters a little older than ours; we always looked forward to seeing them. They lived on a remote mountaintop and had educated all the girls at home.

There were Maryland chair caners, and the pewter makers who made us julep glasses, and the basket weavers whose wares my wife Linda could never resist. Much of my iron and many of my books went in trade to these folks for pieces of their art.

Gradually, I cut out a lot of the smaller, less-known fairs and concentrated on the few well-established ones that were by invitation only — juried shows — and that usually paid honoraria. The emphasis was on demonstration here; the pace was easier; and they were usually at historic locations, like the Lee birthplace, Stratford Hall, or President James Monroe's home, Ash Lawn.

Custom orders were just as good from these shows, and I didn't have to work up and lug a lot of stock to the fairs to sell. We went for the fun of it, the kids exploring, dancing with the entertainers, soaking up good times around the usual first-night barbecue. There was singing and storytelling after a day of tourists, and plans to get together more often before the next fair, even.

A Cautionary Tale

It was at one of these juried shows that I met a gunsmith who was a first-timer there, with his elaborate layout of flintlocks and percussion rifles and pistols. Engraving and inlay were his specialties, and he did handsome work.

I was doing some heat treating — hardening and tempering steel knife blades, chisels, and the like — when he asked me about gun springs. Could I do anything with some springs that were too soft, a whole batch he'd

Craft fairs are good moneymakers. I've just demonstrated a hoop weld here, amid a display of knives, fireplace tools, pot racks, hinges, sconces, and lots of other pieces.

bought cheap, made in Italy? Seems a lot of black powder hardware is manufactured as kits in unlikely places and sold cheap to smiths who mostly assemble them for sale.

Well, you have to understand a little about steel to appreciate the term tempering. As you'll remember, there has to be a certain carbon content to the steel for it to harden in the first place. Then it must be heated, usually to a cherry-red heat, then cooled suddenly, or quenched, for maximum hardness. This is hardening, the first phase.

Next, the steel is reheated slowly with a closely controlled heat, to draw out the hardness, or temper it, to the desired stage. Very hard steel is also very brittle; very soft steel will bend. Most steel items are hardened and tempered somewhere between very hard and very soft, if there is enough carbon in them. If there isn't, nothing happens.

Of course, as you now know, the degree of tempering heat is visible in oxidized colors that change on the surface of the metal. So if you know the carbon content, and have quench hardened the steel from red heat, the colors mean degrees of hardness.

So I tried to explain that I didn't have any way to know whether the gunsmith's Italian springs were recycled tin cans, surgical steel, or old car spring steel. Or if they could be made springier.

Would I try? Well, I guessed so, but it'd take some experimenting. Might break a spring or two finding out just what he had there.

"That's okay. Got 'em cheap. If you can get some stiffness in 'em it'll be worth it."

So I went, first of all, for a hardness that's more or less standard for springs. That is, if spring-steel carbon content is in there. After quenching in oil, I rubbed sandpaper on the spring to shine it, while a crowd gathered and I told, as simply as I could, what I was doing. The gunsmith was all eyes, too.

The first color to appear on the shiny steel is the faint straw, which means a little of the hard brittleness is gone from it. Then the bronze color indicates more heat, and so more hardness gone. Next color is purple, then a clear, peacock blue, which is the usual indicator of spring hardness; again, only if the carbon content is right and you do everything right up till now. Next color would be a gunmetal gray, which means all the hardness is gone and you have soft steel (a relative term).

So I tempered to blue, let the spring cool, and told the gunsmith to try it.

"If that's still too soft, I'll go up a color next time," I told him. "We'll just have to try it one stage at a time." That was okay. He put the spring back in the gun and cocked the hammer.

"Aw, that's just like it was." He was disappointed.

"Okay, give it back here, and I'll do it harder."

"Make it lots harder."

So, again explaining to the crowd that consisted of bored wives and confused husbands, all of whom probably thought it was still black magic, I did it all again. Heated the spring to glowing red, quenched it in oil, shined it with the sandpaper, because it came out of the oil a dirty gray. Then I cranked the forge blower very slowly, moving the spring with my tongs over the red coals.

A straw color came, then bronze. I showed everybody the bronze, which, when you think of it, is a little strange for a piece of plain steel. Alchemists in the Middle Ages probably worked on that a while, trying to make gold.

"Now," I announced, "I'm going to purple, which is one step harder than the blue we tried before, if any of you were here to see that."

"Say," the gunsmith interrupted, "that spring was way too soft. Why don'cha let me try it like it is now? It's lot's harder, ain't it?"

"Well, maybe. We still don't know how much carbon's in there. And if the carbon's right, it's likely too hard. It's up there where it'll probably break if you bend it much."

"Lemme just try it," he reached for it. I don't think any of what I'd tried to explain to him had soaked in. I pulled it out of his reach.

"Okay, but if I don't cool it first, we'll have fried gunsmith here. And this is probably way harder than most springs ever are."

"Don't matter if it breaks. Got it cheap, an' there's lots more."

I had some misgivings. Gun springs are delicate, and there are all sorts of lore about tempering them, most of it colorful if not accurate. You burn a candle under the hardened spring and get it black with soot. Then you heat it till the soot all burns off. Three times. A low chant is suppose to help. So is quenching in ox blood. But I gave him the cooled spring, and he inserted it.

"Feels just right," he announced, flexing the hammer.

"You sure? That's awfully hard for a spring. It could break."

"Nah. That's how I like 'em. Too dead before." He pulled the hammer to half cock. I watched his knuckles whiten, and I knew the spring was too tight.

"Hey, just let me soften that a little. No sense in breaking it if we don't have to. Just take a minute . . ."

"It's okay. Really." To the crowd, "This's just how it's suppose to work, folks. Y'see, the hammer has to come down real hard to strike sparks with the flint. Now I'll just go to full cock."

Everyone watched in fascination. I knew what was coming, but I was the only one. The gunsmith had himself convinced that this piece of steel, so like spaghetti before, couldn't somehow have become brittle enough to break. He hadn't understood a word I'd said. He shifted his hold on the hammer and pulled.

The sound was only a crack, but it snapped heads all over the place. The pewter maker, 50 feet away, jerked around. A tour guide stopped in mid-stride, her flock left-faced.

The gunsmith's face fell as if you'd let down a window shade across it. His look was one of disbelief, then betrayal. He raked me with a look of pure scorn, then spun around and stalked off.

Never spoke to me again.

Making a Go of It

Many smiths double in welding or as mechanics or farriers, just as they would have in 1900. There is seldom enough work to pay rent and overhead for a commercial operation, unless you're a hustler, or have the good fortune to find your niche, and are a good salesman, too.

Steve Stokes, who lives in Albemarle County, Virginia, is a master blacksmith and runs a profitable smithing operation, hiring apprentices, teaching, and catering to the upscale clientele there. His prices are high enough to pay the bills, and his buyers are the folks who appreciate excellent ironwork. He is a third-generation smith, and his Stokes of England shop is becoming widely known. His father not only was a working smith in England, but he also worked with the United Nations leading workshops and writing booklets on the craft.

I often teach courses in blacksmithing. Here, Wayne Haymes and I give instruction to students at the Hollister, Missouri, shop.

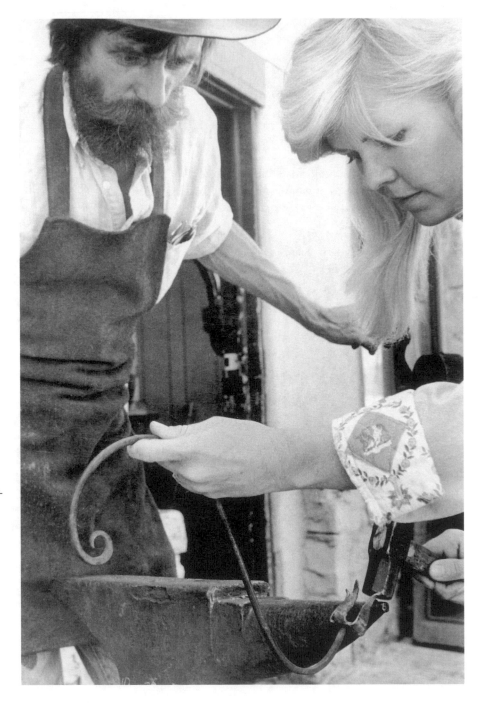

Here, Ginny Ward, a silversmith, puts a finishing curl on the tulip candleholder of a chandelier arm. This was a blacksmithing school project.

But your farmstead shop should be free of overhead beyond your time, a few sacks of coal, and the odd purchases of iron. Let it become known around the rural community that you're hammering out things that can't be bought over any counter, or providing services of a special nature, and you'll have a few jobs soon. A replacement drawbar for an old tractor, a set of harness hooks for a prize team of Clydesdales, a pot rack to fit in an awkward part of the kitchen, an axe retempered, bush-hog blades to be resharpened and heat treated.

You may not make much money, but you'll be able to support your shop easily, and you'll perform a service for the area. And make some friends, because you'll find country people have a liking and respect for anyone who can do things with his hands and isn't afraid to work.

What to Charge

The overriding question regarding selling your work is what to charge for it. Every craftsperson has to work that one out for himself, and not being able to is a big reason a lot of them don't make it.

First of all, you must establish a cost of doing business. Even if your overhead is pennies, you must figure the value of your time, replacement of worn equipment, and cost of additional equipment. All this goes with coal cost, iron cost, transportation, and electricity to determine a shop cost per hour.

Automobile repair shops and welding shops may charge $40 to $75 per hour shop time, to be added to parts costs. Why should your smithy, specializing as it does, charge less? At $20 an hour labor, a project that takes half a day would be $80. Add material cost of, say, $10, since your products are the result mostly of labor.

Four hours is about right for a 36-inch pair of well-made strap hinges, or a fireplace set of poker, shovel, tongs, and stand. Or a pot rack with maybe 20 hooks. If the buyer wants the pieces hand forged, he'll pay the $90 for them. If, down deep, he really doesn't care that much, if he's bargain hunting and would just as soon buy from the discount store, he's at the wrong place. There's another phenomenon at work here, too — the whole has a way of being worth more than the sum of its parts. You can charge more for the set of strap hinges, complete, than the breakdown price. More still if you include a complete Norfolk or Suffolk latch so that the set makes for a completely original door.

If it's important to a buyer that a piece be one-of-a-kind work, it's worth a good price. It's an heirloom. Don't try to compete with the machine age. You don't make any friends, or any profit, by cutting prices. I think I'm typical when I buy handwork — if I see good craftsmanship, price is no longer the question.

Another way to determine price is to set a figure you must be paid for your time. If you can work for $50 a day clear, you might charge only $25 for that four-hour job. Since you won't work every day all day, you'll have to hedge that figure, though — say, double — for the time you can't be at the

Repairing antiques can be a good business. This primitive set of wagon-tire andirons had separated at forge welds in two places.

Triangle dinner gongs sell well and are quickly made of ½-inch mild steel, but they have a sharper ring if they're made of higher carbon stock, such as this one, made of shafting.

shop hammering. More likely, you'll need a lot more, in this 21st century, to keep groceries on the table.

I'll share one more pricing tactic I've developed. Whenever I have something to sell, be it forgework, an old truck, or a piece of equipment (even a house), I set what I consider a fair price. If it doesn't sell in a reasonable time,

A commercial job for another smith. Heating this anvil in order to harden it required most of an afternoon, then it nearly boiled the slack tub dry in quenching. The chain is used to lift it.

I figure I've exhausted the buyers in that price range. You see, regardless of what they say, people buy on price, not quality, usually. So I don't lower the price, looking for a new bunch of prospects, which would probably work (but I'd make less). I raise the price. Gets me that new bunch, who're more convinced of the quality because it costs more. An awful lot of people like to brag about how much they've spent on something unique.

A Woodcarver's Chisel

Blacksmithing is finished on a woodcarver's chisel. The metal collar and ridges will hold a wooden handle in place.

Curving the heated chisel by hammering it against the step of the anvil.

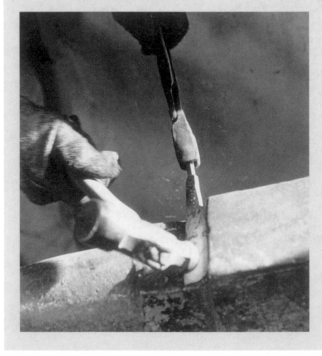

Splitting the end of a wrought-iron fireplace arch brace to make a pot rack. Possibly originally a case-hardened buggy axle, this iron was almost too tough to hacksaw.

A completed pot rack, which will be mounted on a heavy horizontal beam with large lag bolts. Hooks are of ½-inch-square stock, single and split double.

My own time is worth a certain figure to me, arrived at by considering the rate I'd be drawing for consulting work, restoration work, courses, workshops, and writing. I will take a block of that time to make a piece for someone who appreciates it, but only for a comparable rate, plus overhead. It has worked well for me, because I now have enough projects going so that if I don't do the forgework, I'll spend that block of time on one of the other pursuits for income.

I've recently finished a pair of $350 andirons on that basis. No other smith I know would charge that much for a relatively simple (graceful, yes) design, but the owners are ecstatic. They looked at it as buying art, this art being useful, too.

Whatever you charge, be it in dollars, in-kind work, or trades for vegetables and eggs, approach your work at the forge as your specialty, your contribution to the well-being of the rural community. Don't sell it short, and neither will anybody else.

Realize, too, that you are one of a quite select scattering of craftspeople, practitioners of the mighty craft that produced the tools for just about all the rest of them. And be just a little proud.

A commercial job for another smith. This is the upset ball at the end of a vise bar handle. The second ball has to be forged with the vise screw in place.

12

PROSPECTS

Your prospects as a blacksmith today are good, all things considered, whether you plan to set up a commercial operation or not. Even as a hobby, the craft will allow you to create useful and artistic pieces for your own home and shop, as well as for friends. There's a wider appreciation now for things handmade than when we were all in love with the Industrial Revolution.

The people I train who are going on to make their livings from smithing are usually either specializing in ornamental iron or tools, for instance, or combining forge-work with general welding and other metalwork to offer a range of services.

The Advantage of Country Life

Rural areas are excellent for this kind of setup. Neighboring farmers with obsolete equipment can't get replacement parts at the dealership. Often the machine shop or welding shop in town can't or won't make repairs, or their prices are too high. You won't have their overhead, and you'll also be able to create and shape parts these other shops can't. And rural neighbors will patronize you because you're there, once they get to know you. Unless, of course, you're an old grouch.

And when custom orders lag, which they will, you can put your skills to work on such projects as building farm wagons and utility trailers, and repairing worn or broken tools and equipment for resale. The ability to

There will always be a place for custom ironwork, despite the prevalence of commercial, mass-produced tools and decorative ironwork. A client wanted to protect and display his grandfather's tool chest, so he commissioned this hand-forged base. It now holds a place of honor in his home.

use the forge, for one thing, and a knowledge of carbon steels (forging and welding), for another, will make it possible for you to produce quality work cheaply.

A few stock items, such as splitting mauls, digging bars, tow chains, cant hooks, redrawn and tempered pickaxes, and cold chisels, can be kept on hand for sale at a good profit. Fabricated trailers, sleds, and truck beds can be made sturdy with forged bracing, usually with recycled, free iron.

Your advantage over the commercial welding or machine shop in town will be that you're close by, first. And also it's often the case that a part or piece that can't very well be milled or turned or arc welded can be hammered, and maybe hardened and tempered. And your hammering will almost always be far cheaper than the machinist's complicated setup time (he'll be cheaper if he can set up once and make you a hundred parts) or the welder's multiple cuts and welds to produce the same piece (if he can do it at all).

Specializing

Every smith has his specialty. I happen to do more carbon steel work (tools, knives) than anything else, when I have the time. Lately I've been doing more door hinges and latches and light fixtures for the houses we restore. Two of the most successful working blacksmiths I know both do decorative iron, almost exclusively. Another has gone mostly into custom knife work; others are gunsmiths or farriers, having gone on from general working in iron.

Whatever becomes your own specialization, if you do it commercially, you must make clear to your customers that each piece of forgework is the only one there is like that. And that it's going to be around for generations, a sort of heirloom, like other forms of art. And that its price has got to befit such an effort. Your work doesn't compete with the cold-curled, spot-welded, spray-painted twisties of the local carport-post and porch-railing shop, so neither will your prices.

It could get to be a lonely business.

But aside from commercial pursuits, the forge will let you do the thousands of projects on your own place better, cheaper, and faster. One of my recent Virginia students wanted to replace 20 forged shutter latches lost or rusted away on the antebellum house he was restoring. Only a blacksmith could produce those latches, duplicating the remaining originals.

Another former student is always finding things like only one of a set of really unique forged hinges at sales or antiques shops. He's now duplicating these, turning his idle collection into usable hardware for his farm buildings. He's even discovered that a little judicious burning of the iron in the forge makes it look really old — instant rust.

Another man I knew in Missouri operated a sawmill for many years that he and his blacksmith brother built entirely from junkyard parts. Dog hook handles were old piston rods, and the carriage set works ratchet wheel was a cast-off automobile flywheel. Framing was part wood, braced with old straps, shafts, and bars. It was a case of rural necessity, coupled with imagination and that skill with the hammer and the hot iron.

The original and unique pieces you can forge are endless. I've just finished a six-foot pot rack for a friend's new house, made from an old, forged 1780s fireplace arch brace. I matched in design the forged hinges he and I had done for his doors. Such a piece, fitted with single and double pot hooks, could only have been done in the forge by a hand process that did honor to the fine old piece of wrought iron we started with.

Use for Every Project

The value of the forge is made evident every time I teach a workshop. Each student invariably brings to class several old tools or special projects from home to work on. Nothing demonstrates the practicality of the craft better than using these as class projects.

Take a worn pick head, for example. Often a student will bring one that either is too blunt to use or has been drawn out so much there's no metal left. If there's steel to draw, we heat and hammer it out, taking care to work it only within the heat range from cherry red to yellow to protect the carbon steel. Then it's sharpened, hardened, and tempered for use, usually to a bronze for this high-carbon steel.

If the stub of the pick is too little to draw out, we cut and shape a length of heavy coil spring. Then if we're far enough along to get into forge welding, we try a lap weld, though this isn't an easy one to make. Or we might try a shortcut by heating both the new point and the pick head, then arc welding them. The preheating keeps the mass of otherwise cold steel from cooling the weld and metal near it too quickly (almost like a quench) and

When Broadmoor Gardens, near Fort Knox, Kentucky, needed to make a powerful statement, the owner bought a complete iron and stone gate that had been the entrance to a château in France. The journey was not kind to the gate — design elements were broken; the stanchions into the ground were gone; and the connectors were bent, broken, or missing. My crew and I were commissioned to repair the gate, prepare the site along the highway, and erect all the elements for the 50-foot-long, 21-foot-high entryway. My blacksmithing skills — forge welding, electric welding, building the replacement ornamental details at the forge, repairing the existing design elements — were what made the job possible. Today, people know they have arrived at Broadmoor Gardens when they come upon this incredible gateway.

keeps it from cracking. Then we harden and temper normally. The value of this exercise is that it utilizes several smithing procedures in one useful project.

I never have students do "make-work" projects. Everything has its use. Even a forged chain link will perhaps join a hammered hook to a piece of chain found along the roadside or bought in a box of odds and ends at a flea market.

At any given time, I can list a dozen or so forged pieces I've done within, say, the past week for my place. At this writing they'd be:

A slender, sharp punch for prepunching nail holes in the sheet-metal roof for my garage. A new bar handle for the vise I traded for. Braces for the wooden sides of my utility trailer. Pothooks for cast-iron cookware in our kitchen. A new machete for hacking the briars and honeysuckle jungle. A miniature cleaver for my wife to chop vegetables with. New round-hole punches for use in the smithy itself. A pair of footrests for the antique wood-

Over the years, my daughter, Lauren, has gone into the blacksmith shop and made several items she needed for her post-college apartments. These have included bathroom-paper holders, paper-towel holders, dinner gongs, hooks, and this boot scraper.

My son, Charlie, made this wood rack as a Christmas gift for his sister's new digs, which she heated with a small woodstove.

burner we heat with. A new fire shovel so I won't have to borrow the coal shovel from the shop anymore. A basket-twist handle for a friend's antique cast-iron teakettle in trade for the use of his air compressor and sandblaster to get ahead of the rust on an antique car of mine, for which I'll soon forge new bumper braces and other irreplaceable parts.

Smithing Without Boundaries

For many modern smiths, the craft is simply a fascinating step beyond the woodworking shop or the garage fix-it bench. The difference, and most of the mystique, is that there are almost no limits to the smithy. You can actually forge the tools for any other craft I can think of, even the wrenches for fixing a sick engine, if you want to go that far. Or even (are you ready for this?) hammer out an entire engine itself. (How do you think the first ones were made?)

Just how far you go with your hammer in hand is up to no one but you. As your skill increases, you'll want to tackle more complex problems. Can you really build a pulled road scraper for that half-mile of potholes to your place? Sure. Use an old pickup axle or two and a cutout pipeline section for the blade, edged with a cast-off grader cutting edge, and gears, racks, and pinions from the junkyard.

Or make your own woodstove from maybe a small, test-rejected pressure tank, with forged fire door hinges and scrolled feet (my friend Al Lemons has done several of these). Or your own plow or spike harrow or rock rake or earth-moving scoop. Or your own water wheel for a power generator or irrigation pump down on the creek.

You see, until this age of punched-out, stamped, assembly-lined, robot-assembled, obsolescence-riddled throwaways, everything iron was made in the forge.

It still can be.

Never Let Them Call You a "Smithy"

A final note — don't let the assorted name-dropping sages get away with calling you "Smithy." (It's like being called a garage.) Everybody seems able to remember and misinterpret the first line from Longfellow's poem "The Village Blacksmith" — "Under the spreading chestnut tree, the village smithy stands." But they never can remember the second line — "The smith, a mighty man is he, with arms like iron bands." But at 145 pounds wringing wet, somehow I hardly fit the traditional picture.

GLOSSARY

Acetylene. Gas used with oxygen to weld or cut steel/iron.

Andirons. Metal supports for logs in a fireplace.

Anneal. To soften metal; steel is softened by heating then cooling very slowly.

Anvil. Shaped surface of heavy steel on which hot metal is hammered.

Arc weld. Electric welding method to fuse two metals.

Ash dump. Device for cleaning ashes from the forge.

Auger. Boring bit for making holes in wood.

Babbitt. Soft metal used in bearings.

Ball-peen hammer. Hammer with rounded-back face.

Barking spud. Tool for peeling bark from logs.

Bearing. Device allowing free turning of a shaft or wheel.

Bellows. Flexible, valved enclosure for forcing air into the forge.

Blacksmithing. Working iron or steel.

Boomer. Chain binder or tightener.

Borax. Substance used as welding flux to seal out oxygen, which forms scale.

Brazing. Joining with molten bronze or brass.

Brine. Salt water, often used to quench and harden hot steel.

Butt weld. Joint made by forging two white-hot ends together.

Cant hook. Handled hook for moving logs.

Carbon content. Amount or degree of carbon in steel.

Carbon steel. Steel that has enough carbon content to harden for use as tools.

Case hardening. Making steel out of iron by letting it absorb carbon on the surface.

Chain binder. Device for hooking and tightening chain.

Charcoal. Partly burned wood, sometimes used for forge fuel.

Clinch cutter. Tool for cutting nails, as in horseshoeing.

Clinker breaker. Device for clearing the forge fire of clinkers.

Clinkers. Melted-out impurities in coal.

Coil steel. Usually, coil springs.

Coke. Partly burned coal or petroleum, used as fuel in blacksmithing.

Cold chisel. Chisel used for cutting cold metal.

Countersink. To recess a hole.

Cross-peen hammer. Hammer with a blunt chisel back face 90 degrees to the handle.

Crystalline structure in metal. The pattern of particles that comprise a metal.

Cutting torch. Acetylene or LP gas/oxygen torch for cutting iron/steel.

Damascus steel. Pattern-welded, layered wrought iron and carbon steel.

Decalescence (point of). Temperature at which steel molecules are in a state of flux, allowing hardening by quenching or sudden cooling.

Die. Tool for cutting external screw or bolt threads.

Digging bar. Chisel-pointed steel bar.

"Draw a piece of iron." To flatten, extend by heating and hammering; to stretch.

Drawknife. Two-handled tool for shaving wood.

"Draw to a . . ." Either to flatten and extend metal by hammering or to heat to a given tempering color.

Drift out. To enlarge, as a hole.

Drifts. Tools for enlarging holes by punching.

Drill press. Stationary drilling machine.

Farrier. One who applies iron shoes to horses' feet.

Farrier vs. blacksmith. Farrier shoes horses, blacksmith works iron/steel; sometimes one does both.

Firebrick. Brick that withstands heat.

Firepot. Shaped center of the forge into which air is forced.

Flatter. Struck tool with handle, used to produce a flat, controlled surface on hot iron.

Flue. Chimney stack.

Flux. Substance, usually containing borax, to retard oxidation of heated iron/steel.

Forge. Container for fire to heat metal.

Forge weld. To join iron/steel by hammering while white hot.

Foundry. Operation for casting iron/steel.

Froe. Tool for splitting (riving) shakes or shingles.

Fuller. Two-piece tool, the bottom like a blunt hardy, the top like a blunt cold chisel, with or without a handle, used to shoulder or dent hot iron with more accuracy than the hammer peen.

Gas forge. Forge fed by gas and forced air.

Grain in metals. The elongated particles that make up metal.

Grate. Device for holding fire, to let ashes fall through.

Hardy. Stationary chisel set into the hardy hole in an anvil.

Hardy hole. Square hole in the face of an anvil for holding tools.

Heat treating. Combination of hardening and tempering steels and other metals.

"Heeling the hammer." Holding the handle at a low angle.

Hold an edge. Resist dulling.

Hone. Abrasive tool for sharpening.

Horn of the anvil. Rounded extension, used for working loops, curves.

Hot cut. Chisel used on hot metal.

"Hot short." Tendency of some iron/steel to crumble when worked.

Iron. Relatively soft ferrous metal containing no carbon.

Jump weld. Forge weld made by butting the end of one piece into the side of another and hammering at white heat.

Lap scarf. The shape given ends of metal to be welded in an overlap.

Leaf spring. Vehicle spring comprised of layers.

Low sulfur. In coal, that containing little sulfur.

Machinist's vise. Vise that mounts on a tabletop.

Masonry forge. Forge set into brick or stone.

Maul. Large hammer, or chisel-pointed hammer for splitting wood.

Metallurgy. The study of metals.

Mild steel. That with little or no carbon content, which cannot be hardened.

Mill pick. Tool for dressing millstones.

Molten. Liquid state, as hot iron/steel.

Mushrooming. The spreading of a struck surface of a tool.

Normalize. To cool a heated metal in the air to its normal hardness.

Offset. Often, a double bend to create parallel but nonaligned sections of a shaft.

Oxidation scale. Thin flakes of oxidized metal, caused by heat.

Oxidize. To combine with oxygen, which produces scale in iron/steel.

Pan forge. A shallow forge with no firepot.

Peavy. Similar to a cant hook; a tool for handling logs.

Peen. Shaped face of a hammer (ball, cross, angled, straight).

Peter Wright. Patent anvil brand, English.

Pintle. An anchored pin from which to hinge a door or gate.

Pitting. Cavities in the metal caused by burning or oxidation.

Pivot bolt. A bolt that holds, but allows swiveling.

Pritchel hole. The round punch (pritchel) hole in the anvil face.

Prybar. Also pinch bar or wrecking bar; bar with a hook and split chisel point at one end, straight with chisel at the other.

Punch. Tool for making holes in hot metal or for pushing objects through holes.

Purlins. Timbers parallel to the ridge in a roof.

Pyrometer. High-temperature thermometer.

Quarry steel. That used for stone cutting.

Quench. To harden by sudden immersion in liquid.

Ratchet blower. A blower driven by repeated engagement of a one-way, sliding lever.

Rebar. Construction slang for reinforcing rod.

"Red short." Same as "hot short"; tendency of some iron/steel to crumble when worked.

Reface (an anvil). To resurface the top by welding on a plate of carbon steel.

Rive. To split, as in wooden shakes or shingles.

Rivet. To join pieces by flattening the ends of a short rod through holes in them.

Scale. Oxidized metal that forms in thin sheets from heat.

Scarfing. Shaping ends to be welded.

Scarf weld. Weld for which the components have been shaped.

Shear. In blacksmithing, a tool for cutting iron/steel.

Slack tub. Originally "slake"; a tub for quenching or cooling hot metal.

Sledge (sled). Simple vehicle on runners for moving material.

Smith. The craftsperson who works metals.

Smithy. The shop in which the smith works.

Splitting maul. Chisel-pointed hammer for splitting wood.

Spring steel. Steel of high carbon content used for springs.

Steel. Iron with carbon in it.

"Steel an axe." To weld new steel onto the cutting edge.

Straight-peen hammer. Hammer with the blunt chisel back face parallel to the handle.

Sulfur. Substance found in coal.

Swage. Two-piece tool, the bottom fits the anvil hardy hole, the top is handled; each has a shaped, matched groove in which iron is shaped while hot.

Swage block. Heavy cast-iron block with shaped holes and grooves for shaping hot iron.

Tang. Extended end of a tool for attaching a handle.

Tap and die. Nut-and-bolt threading tools.

Temper. To regulate hardness in metal by controlled heating, after hardening.

Temper drawing. Tempering; taking quench hardness out to a desired degree by controlled heat; the second step in heat treating.

Tempering fluids. Those used in quenching; usually water, brine, or oil.

"Toeing the hammer." Angling the handle too high.

Tongs. Gripping tool for holding hot metal.

Tool steel. Steel of high carbon content used for tools to work other steel cold or to work stone.

Tough steel vs. brittle (hard) steel. Medium-carbon steel vs. high-carbon steel; the ability of the former to withstand breakage, the latter to withstand wear.

Trip-hammer. Powered hammer/anvil machine.

Trunnel. "Tree nail"; wooden peg used to hold timbers.

Tuyere. Shaped air blast into forge firepot.

Upset. To thicken by hammering into a mass, as opposed to stretching (drawing out).

Vise (stump or tree). Also blacksmith's vise or leg vise; holding device with a long leg anchored below for extra stability.

Whetstone. Stone for sharpening tools.

Wrought iron. Soft iron and silica, very fibrous, now rare.

SUGGESTED READING

Andrews, Jack. *New Edge of the Anvil*. New York: SkipJack Press, 1994.

Bealer, Alex W. *The Art of Blacksmithing*. Edison, New Jersey: Book Sales, 1995.

Meilach, Dona Z. *Decorative and Sculptural Ironwork*. Atglen, Pennsylvania: Schiffer Publishing, 1999.

Richardson, M. T. *Practical Blacksmithing,* volumes 1 and 2. Mendham, New Jersey: Astragal Press, 1998.

Watson, Aldren A. *The Village Blacksmith*. New York: T.Y. Crowell, 1977.

Weygers, Alexander G. *The Complete Modern Blacksmith*. Berkeley, California: Ten Speed Press, 1997.

INDEX

Page numbers in *italics* indicate photos or illustrations. Page numbers in **bold** indicate tables.